The Wastewater Operator's Guide to Preparing for the Certification Examination

2016

Water Environment Federation
601 Wythe Street
Alexandria, VA 22314-1994
www.wef.org

Association of Boards
of Certification
http://www.abccert.org
abc@abccert.org

Certification Commission for
Environmental Professionals
http://www.professionaloperator.org
info@professionaloperator.org

IMPORTANT NOTICE

WATER ENVIRONMENT FEDERATION

The Water Environment Federation (WEF) is a not-for-profit technical and educational organization of 33,000 individual members and 75 affiliated Member Associations representing water quality professionals around the world. Since 1928, WEF and its members have protected public health and the environment. As a global water sector leader, our mission is to connect water professionals; enrich the expertise of water professionals; increase the awareness of the impact and value of water; and provide a platform for water sector innovation. To learn more, visit www.wef.org.

ASSOCIATION OF BOARDS OF CERTIFICATION

The Association of Boards of Certification (ABC) was founded with the mission to advance the quality and integrity of environmental certification programs throughout the world. This charge has held strong through more than 40 years of providing knowledge and resources to nearly 100 certifying authorities representing more than 40 states, 10 Canadian provinces and territories, and several international and tribal programs. ABC believes in certification as a means of promoting public health and the environment while striving to give our members the necessary tools to ensure the knowledge and skills of their operators.

CERTIFICATION COMMISSION FOR ENVIRONMENTAL PROFESSIONALS

ABC established the Certification Commission for Environmental Professionals (C2EP) in 2012 as an independent entity of water environment industry and certification subject matter experts dedicated to enhancing the integrity and standardization of operator certification. C2EP has been instrumental in the creation of the Professional Operator (PO) certification and designation—the industry's first and only professional designation recognized internationally.

By earning Professional Operator certification in water treatment, water distribution, wastewater treatment, or wastewater collection system operations, operators demonstrate their ability to meet international standards and their desire to grow as professionals. This, along with the mandatory adherence to the Professional Operator Code of Conduct, promotes public visibility, consumer confidence, and support of water professionals and the industry.

The Association of Boards of Certification and C2EP have identified the following WEF manuals as primary reference material to help operators prepare for the ABC/C2EP wastewater operator certification exams:

- *Operation of Municipal Wastewater Treatment Plants*—Manual of Practice No. 11, 6th Edition, and
- *Activated Sludge*—Manual of Practice No. OM-9, 2nd Edition.

In addition to the primary references listed above, WEF has many other technical resources that are listed in the study guide. If you are having difficulty answering any of the questions in the study guide, you should consult the reference source provided following the answer to the question.

For information on PO certification and professional designation, testing services, or other ABC/C2EP opportunities, contact

Association of Boards of Certification
http://www.abccert.org
abc@abccert.org

Certification Commission for Environmental Professionals
http://www.professionaloperator.org
info@professionaloperator.org

1-515-232-3623
2805 SW Snyder Blvd., Suite 535
Ankeny, IA 50023 USA

CONTENTS

PREFACE

The Association of Boards of Certification (ABC), the Certification Commission for Environmental Professionals (C2EP), and the Water Environment Federation (WEF) have worked together to produce this document based on industry-wide job analyses and *Need-to-Know Criteria*. This publication is designed to give users practice answering multiple-choice exam questions that will be similar in format and content to those they might find on the certification examination.

The questions test the skills and knowledge required of an operator working in a water resource recovery facility. Some of the questions are more difficult and complex than others.

WEF/ABC/C2EP Disclaimer

The questions included in this guide have been chosen to sample as many different aspects of a wastewater operator's job responsibilities as possible. However, because of the tremendous variety in equipment, processes, conditions, and operator duties, not all of the questions may be useful in all of the possible certification applications. It should also be noted that questions in the study guide are only intended to provide an example of style and possible topics for certification exam questions and will not be found on an ABC/C2EP certification exam.

ACKNOWLEDGMENTS

The following individuals participated in the development of *The Wastewater Operator's Guide to Preparing for the Certification Examination:*

Task Force Chair, Sidney Innerebner, Ph.D., P.E., C.W.P., Indigo Water Group, LLC, Littleton, Colorado

Authors:

Geraldine L. Ahrens, Pahrump, Nevada
Pitiporn Asvapathanagul, Norwalk, California
Michael S. Beattie, Brown & Caldwell, Saint Paul, Minnesota
Paul Burris, Manteno, Illinois
Richard E. Finger
Michael T. Fritschi, South Suburban Sanitary District, Klamath Falls, Oregon
Francis J. Hopcroft, Wentworth Institute of Technology, Boston, Massachusetts
Sidney Innerebner, Ph.D., P.E., C.W.P., Indigo Water Group, LLC, Littleton, Colorado

Paul Krauth, P.E., Utah Division of Water Quality, Salt Lake City, Utah
Christopher Kuhlemeier, Clark County Water Reclamation District, Las Vegas, Nevada
Jorj Long
John Meyer
Stacy J. Passaro, P.E., BCEE, Passaro Engineering LLC, Mount Airy, Maryland
Kim R. Riddell, Alloway
John Saturley, CPEA, Littleton/Englewood Wastewater Treatment Plant, Littleton, Colorado
Kenneth Schnaars, P.E., Brown and Caldwell, Nashville, Tennessee
Eric J. Wahlberg, Brown and Caldwell, Irvine, California
David Wright, CSHO, Weston&Sampson, Woburn, Massachusetts

Reviewers:

Ata Adeel, E.I., C.W.P., Ohio Regional Sewer District
Charles "Chuck" Corley, former Water Pollution Control Division Manager, Illinois Environmental Protection Agency, Rockford, Illinois
Frank DeOrio, Technical Director, O'Brien & Gere Engineers, LLC, Syracuse, New York
David Gray, Chief Operator, Surfside Wastewater Facility, Nantucket Public Works, Nantucket, Massachusetts
Chris Hoffman, Research Scientist 1, Division of Water Quality, New Jersey Department of Environmental Protection, Trenton, New Jersey
Samuel S. Jeyanayagam, Ph.D., P.E., BCEE, *Chair*
Paul Krauth, P.E., Outreach Coordinator, Utah Division of Water Quality, Salt Lake City, Utah
Dave Mason, Kingwood, Texas
Andrew O'Neill, Wastewater Technical Outreach Operator, Washington State Department of Ecology, Spokane, Washington
John Reynolds, PO, former Senior Wastewater Treatment and Collections Operator, Sooke Wastewater Treatment Plant, Sooke, British Columbia
LeAnna Risso, Assistant Manager of Process Control, AAS Environmental Safety & Health, Clark County Water Reclamation District, Las Vegas, Nevada
John "Jack" Vanderland, former Director of Wastewater Operator Training and Assistance Program, Virginia Department of Environmental Quality, Richmond, Virginia

CERTIFICATION COMMISSION FOR ENVIRONMENTAL PROFESSIONALS EDUCATION AND EXPERIENCE REQUIREMENTS

The Certification Commission for Environmental Professionals (C2EP) of ABC offers certification and professional designation to wastewater treatment operators through its Professional Operator certification program. In addition to passing an exam and agreeing to the Professional Operator Code of Conduct, successful candidates must also satisfy education and experience requirements to earn Professional Operator certification and professional designation. Although a number of education and experience substitutions are allowed, basic formulations of the eligibility criteria for each class of certification follow:

CLASS I

- High school diploma, general equivalency diploma (GED), or equivalent;
- One year of acceptable operating experience; and
- Ninety contact hours of postsecondary education.

CLASS II

- High school diploma, GED, or equivalent;
- Three years of acceptable experience; and
- One hundred eighty contact hours of postsecondary education.

CLASS III

- High school diploma, GED, or equivalent;
- Four years of acceptable operating experience, including 2 years of direct responsible charge; and
- Nine hundred contact hours of postsecondary education.

CLASS IV

- High school diploma, GED, or equivalent;
- Four years of acceptable operating experience including 2 years of direct responsible charge; and
- One thousand eight hundred contact hours of postsecondary education.

ABC and C2EP also provide testing and certification services to many states and provinces. In these instances, the individual states, provinces, or other agencies have their own eligibility criteria and/or exam content that likely differ in some manner. Contact your certification authority directly to receive the latest information regarding their specific eligibility requirements and exam specifications.

ASSOCIATION OF BOARDS OF CERTIFICATION WASTEWATER TREATMENT EXAM SPECIFICATIONS

Four levels of certification exams are offered by ABC, with Class I being the lowest level and Class IV the highest level. The exam specifications are based on a weighting of the job analysis results so that they reflect the significance of tasks performed on the job. The specifications detail the number of calculation items included on each exam and list the percentage of questions on the exam that fall under each content domain of related job duties. For example, 25% of the questions on the ABC Class I wastewater treatment exam relate to the content domain "Equipment Operation".

Job Duties and Complexity

Specific testable job duties associated with each of the content domains are listed in the *Need-to-Know Criteria* preceding the practice questions for that domain. Some of these job tasks are more simple and routine, whereas others are more complex or mentally demanding. For each content domain, the exam specifications detail the number of exam items that will fall under each of three levels of complexity: recall, application, and analysis.

- **Recall** items typically just require the simple recall or recognition of specific facts, concepts, processes, or procedures, with little to no problem-solving involved.
- **Application** items will involve some basic problem-solving, calculations, or the interpretation and application of data.
- **Analysis** items may involve higher level problem solving, evaluation, or the fitting together of a variety of elements into a meaningful whole, as in inductive reasoning. Tasks at this level will typically require many steps in the thought process.

Related Knowledge

Listings of the types of underlying knowledge that support the performance of the job duties included in a content domain are provided alongside the Need-to-Know Criteria preceding the practice questions for that domain. Both the types of knowledge within each content domain and the level of that knowledge vary based on examination class. The complexity of the required knowledge increases with increasing job complexity. A three-level taxonomy of basic, intermediate, and advanced is used to rate the extent of knowledge needed to perform the tasks assigned to each content domain.

- **Basic**—A fundamental, or lower level of knowledge is required. Operators performing tasks requiring this level of knowledge will be able to do so with some training. This level of knowledge may also be acquired and developed through job experience. Not having this level of knowledge will have minimal effect or significance on the performance of the tasks listed in the content domains, or on public safety and welfare.

Performing tasks requiring this level of knowledge may be routine, using established procedures, and have a low level of complexity. The operator will not be required to fully understand and discuss the application and implications of changes to processes, policies, and procedures within the content domain. The operator will be performing directly assigned tasks and will work under immediate supervision.

- **Intermediate**—A level of knowledge beyond the basic level is required. Operators performing tasks requiring this level of knowledge will be able to do so with training beyond that of the basic level. The intermediate level will require mastery of the knowledge required at the basic level and will also require substantial knowledge beyond that level. Not having this level knowledge will have a significant effect on the performance of the job and on public safety and welfare.

  The operator will not only be able to apply required fundamental concepts, but will be able to understand and discuss the application and implications of changes to processes, policies, and procedures within the content domain. Work can be performed under limited supervision and may be more varied and demanding, while also requiring some degree of responsibility. Although assistance will sometimes be required from someone holding an advanced, or expert level of knowledge, tasks can typically be performed independently with an intermediate level of knowledge.

- **Advanced**—A very high level of knowledge/job expertise is required. The advanced level will require mastery of the knowledge required at the basic and intermediate levels. Operators performing tasks requiring the advanced level of knowledge will possess the training and skills well beyond that of the intermediate level, and will be functioning at an expert level. Not having this level knowledge will have a serious effect on the performance of the job and will be harmful to public safety and welfare.

  The operator can apply all fundamental and highly developed or complex concepts and will be able to design, review, and evaluate processes, policies, and procedures within the content domain. Job-related decision-making can be performed with little to no supervision. The operator with an advanced level of knowledge can operate with a great deal of autonomy and provide guidance and training for operators with basic and intermediate levels of knowledge.

Class I	Total Number of Items	Number of Items by Complexity			Number of Calculation Items
		Recall	Application	Analysis	
Laboratory Analysis	10	2	7	1	
Equipment Evaluation and Maintenance	25	12	10	3	
Equipment Operation	25	8	14	3	
Treatment Process Monitoring, Evaluation, and Adjustment	30	5	10	15	10
Security, Safety, and Administrative Procedures	10	7	3	0	
Totals	**100**	**34**	**44**	**22**	

Class II	Total Number of Items	Number of Items by Complexity			Number of Calculation Items
		Recall	Application	Analysis	
Laboratory Analysis	15	3	9	3	
Equipment Evaluation and Maintenance	20	8	9	3	
Equipment Operation	25	5	15	5	
Treatment Process Monitoring, Evaluation, and Adjustment	30	5	10	15	12
Security, Safety, and Administrative Procedures	10	6	4	0	
Totals	**100**	**27**	**47**	**26**	

Class III	Total Number of Items	Number of Items by Complexity			Number of Calculation Items
		Recall	Application	Analysis	
Laboratory Analysis	15	2	8	5	
Equipment Evaluation and Maintenance	20	6	11	3	
Equipment Operation	25	5	10	10	
Treatment Process Monitoring, Evaluation, and Adjustment	30	5	10	15	15
Security, Safety, and Administrative Procedures	10	4	6	0	
Totals	**100**	**22**	**45**	**33**	

Class IV	Total Number of Items	Number of Items by Complexity			Number of Calculation Items
		Recall	Application	Analysis	
Laboratory Analysis	20	3	10	7	
Equipment Evaluation and Maintenance	15	4	6	5	
Equipment Operation	20	2	10	8	
Treatment Process Monitoring, Evaluation, and Adjustment	35	5	15	15	18
Security, Safety, and Administrative Procedures	10	2	4	4	
Totals	**100**	**16**	**45**	**39**	

IMPORTANCE OF CERTIFICATION

The wastewater treatment operator is responsible for taking quick and effective action to protect public health and the environment by ensuring that wastewater treatment infrastructure is operating properly and that this infrastructure is properly maintained and protected to achieve a maximum usable life.

Today's wastewater treatment operators must understand how each component of a system functions and how these components come together to treat incoming waste streams. Properly operating and maintaining a wastewater treatment system requires knowledge and competency in a broad range of topics and skills. Certification provides operators with a third-party-verified yardstick of achievement that demonstrates their level of accumulated knowledge, skills, and experience that are important in the day-to-day activities their jobs require.

Many utility owners and managers place value and importance on an operator's steady progress through the various levels of certification. Indeed, it demonstrates a focus and commitment to building knowledge and skills that improve decision-making and actions on the job. Achieving a high level of certification allows an operator to stand out among his/her peers. It also can help to show that an operator is ready to move to a higher level within an organization. It also can help a new utility assess the achievement of an operator they are interviewing. It is a professional title that an operator will take with him/her throughout his/her career.

TAKING THE CERTIFICATION EXAMINATION

Here are some helpful tips for successfully taking a certification examination.

BE PREPARED

The first step in passing a certification examination is to prepare for it in advance by study-ing the type of information that will be covered on the examination. One method of pre-paring for the examination is to attend training classes offered by local wastewater utilities, community colleges, and vocational–technical schools. Another useful method is to read and study reference books on wastewater treatment. Many state certification agencies pub-lish or can recommend reference materials that can help you study the material covered on the examination in your state. In addition, some state or provincial environmental depart-ments provide sample tests, lists of material covered in the examinations, or examples of typical test problems. Each question in this guide is referenced to a technical source. If you find you are having difficulty with a specific topic, you may want to review the material in that reference.

If you are preparing for the examination on your own, it is critical that you work out a doable schedule that will allow you ample time to work through each of the important topic areas that the examination will cover. To do this, review each of the core competencies and decide what material you want to review for each and how many sample questions you will take for each competency. Determine how many hours of study time will be required to complete the review and practice questions. On your schedule, working back from the date when you would like to take the examination, schedule blocks of study time on specific days for each topic for the total amount of time required. Make sure you stick with this schedule as you prepare for the examination. Do not "cram" or try to study all the material during the week or so before the examination. This is not an effective way to study.

Certification examinations are intended to test every aspect of an operator's involve-ment in wastewater treatment system operation and maintenance. Because calculations are often involved, certification examinations include mathematical problems dealing with flowrates, pump performance, wet well volumes, and so forth. To prepare for the mathe-matical problems, it is a good idea to memorize the more common conversion factors (con-stants) used in wastewater treatment system mathematics. Formula/conversion tables are handed out at ABC/C2EP examinations. An abbreviated version of this is included at the end of this section.

Remember that examinations are changed frequently and computers can be used to generate random exams from larger sets of questions. Therefore, do not try to learn the answers to specific problems that you may have heard about from other operators. Con-centrate on learning how to work various types of problems, such as how to calculate a target return sludge rate or how to adjust a chemical feed pump. This way, you will be prepared to handle any question on the examination, not just certain questions you have memorized.

Although operators are responsible for knowing all aspects of wastewater treatment, certification examinations in different regions emphasize some aspects more than others. This guide is broken into topics that are commonly emphasized on the examinations. By talking with your certification program coordinator, you can learn which additional areas may be emphasized on the examination given in your area.

As a final tip for preparing for the examination, give yourself the night before the examination off from studying. Do something relaxing and try to get to bed early and get a good night's rest. Again, do not try to "cram" information into your brain during the last few hours. Cramming typically is not beneficial and can lead to confusion and exhaustion.

TAKING THE EXAMINATION

On the day of the examination, arrive at the testing location well ahead of the announced starting time. Be sure that you have the necessary materials, such as a spare pencil and your admittance slip (if required). Also, remember to put fresh batteries in your calculator because working mathematical calculations by hand can waste valuable time. Before starting the examination, skim through the questions to get a general idea of the kinds of questions that are on the examination. By seeing the types of questions and math problems you must answer at the start, you will have a better idea of how to use your time. It is a good idea to answer all the easy questions first and then tackle the more complex problems that will take longer to complete.

If the examination has essay questions and lengthy math problems, think about the person who will grade these questions. To help the grader, write your answer clearly and in an order that the grader can easily follow. It is important to remember that graders are human and that they will be reading lots of essays and math problems. Although they may not intend to lower your grade for sloppy or unreadable work, such work certainly will not help your score. Underline answers to math problems; do not make the grader search through pages of figures to find your answer.

CALCULATION PROBLEMS

Calculation problems are an important and often difficult part of certification examinations. Follow these guidelines when solving calculation problems:

(1)   Read each question carefully to ensure that you know what answer is required;

(2)   Make a drawing or sketch if it will help you solve the problem;

(3)   Simplify the problem. If the problem is complex, break it down into small pieces that you can solve separately;

(4)   Make all necessary conversions (such as converting 1.5 ft to 18 in., if the question requires an answer in inches);

(5)   Give your answer only to the nearest significant figure. For example, a chemical dosage problem with an answer of 26.93276 lb is meaningless when the chemical feeder is only accurate to the nearest pound;

(6)   Be sure the decimal point is in the right place; and

(7)   Check to see that your answer makes sense. For example, if you get an answer of $30,000,000,000 for a question that asks for the annual cost of odor control chemicals for a 0.5-mgd pumping station, your answer is obviously wrong. Try to find another approach to solving this math problem.

HELPFUL HINTS

Most certification examinations are primarily composed of multiple-choice questions. You should never leave a multiple-choice question blank. Often, you can determine that one or two of the potential answers are obviously wrong. If you narrow your choice to only two possible answers, you have a 50/50 chance of getting the right answer, no matter which answer you choose.

Take the examination methodically and deliberately. Avoid the natural temptation to rush through it. Be relaxed while taking the examination. Periodically, take a moment while sitting in your seat to stretch your neck and shoulders to ease tension.

EXAMINATION CHECKLIST

Before leaving your home or work to go to the testing location, be sure that you have the following items so that you will be well prepared to take the examination:

- An admittance slip (if one is required),
- Two sharpened no. 2 pencils with good erasers,
- A calculator,
- Fresh batteries for the calculator,
- A watch,
- Glasses (if you wear them), and
- Confidence.

Remember, if you have properly prepared for the examination, you should do well. Stay calm and carefully work your way through the questions.

Best of luck with your examination and your career as a wastewater treatment operations professional!

Formula/Conversion Table for Wastewater Treatment, Industrial, Collection, and Laboratory Exams in SI Units

Alkalinity, as mg CaCO$_3$/L $= \dfrac{\text{(Titrant Volume, mL)(Acid Normality)(50 000 mg/equiv)}}{\text{Sample Volume, mL}}$

Amps $= \dfrac{\text{Volts}}{\text{Ohms}}$

Area of Circle $= (0.785)(\text{Diameter}^2)$ or $= (\pi)(\text{Radius}^2)$

Area of Cone (lateral area) $= (\pi)(\text{Radius})\sqrt{\text{Radius}^2 + \text{Height}^2}$

Area of Cone (total surface area) $= (\pi)(\text{Radius})(\text{Radius} + \sqrt{\text{Radius}^2 + \text{Height}^2})$

Area of Cylinder (total outside surface area) $=$ [Surface Area of End #1] + [Surface Area of End #2] + [(π)(Diameter)(Height or Depth)]

Area of Rectangle $=$ (Length)(Width)

Area of Right Triangle $= \dfrac{\text{(Base)(Height)}}{2}$

Average (arithmetic mean) $= \dfrac{\text{Sum of All Terms}}{\text{Number of Terms}}$

Average (geometric mean) $= [(X_1)(X_2)(X_3)(X_4)(X_n)]^{1/n}$ The nth root of the product of n numbers

Biochemical Oxygen Demand (unseeded), mg/L $= \dfrac{[(\text{Initial Dissolved Oxygen, mg/L}) - (\text{Final Dissolved Oxygen, mg/L})](\text{Sample Volume, mL})}{\text{Diluted Volume, mL}}$

Chemical Feed Pump Setting, % Stroke $= \dfrac{\text{(Desired Flow)(100\%)}}{\text{Maximum flow}}$

Chemical Feed Rate, mL/min $= \dfrac{(\text{Flow, m}^3/\text{d})(\text{Dose, mg/L})}{(\text{Chemical Feed Density, g/cm}^3)(\text{Active Chemical, \%})(1440)}$

Circumference of Circle $= (\pi)(\text{Diameter})$

Composite Sample Single Portion $= \dfrac{\text{(Instantaneous Flow)(Total Sample Volume)}}{\text{(Number of Portions)(Average Flow)}}$

$$\text{Cycle Time, minutes} = \frac{\text{Storage Volume, m}^3}{\text{Pump Capacity, m}^3/\text{min} - \text{Wet Well Inflow, m}^3/\text{min})}$$

$$\text{Degrees Celsius} = [(\text{Degrees Fahrenheit} - 32)(\tfrac{5}{9})] \text{ or } \frac{(^\circ F - 32)}{1.8}$$

$$\text{Degrees Fahrenheit} = [(\text{Degrees Celsius})(\tfrac{9}{5}) + 32] \text{ or } [(\text{Degrees Celsius})(1.8) + 32]$$

$$\text{Detention time} = \frac{\text{Volume}}{\text{Flow}} \qquad \text{Note: Units must be compatible.}$$

$$\text{Dose} = \text{Demand} + \text{Residual}$$

$$\text{Electromotive Force (E.M.F.), volts} = (\text{Current, amps})(\text{Resistance, ohms}) \text{ or } E = IR$$

$$\text{Feed Rate, kg/d} = \frac{(\text{Dosage, mg/L})(\text{Flowrate, m}^3/\text{d})}{(\text{Purity, Decimal Percentage})(1000)}$$

$$\text{Filter Backwash Rate, L/m}^2 \cdot \text{s} = \frac{\text{Flow, L/sec}}{\text{Filter Area, m}^2}$$

$$\text{Filter Backwash Rise Rate, cm/min} = \frac{\text{Water Rise, cm}}{\text{Time, minute}}$$

$$\text{Filter Yield, kg/m}^2 \cdot \text{h} = \frac{(\text{Solids Concentration, \%})(\text{Sludge Feed Rate, L/h})(10)}{(\text{Surface Area of Filter, m}^2)}$$

$$\text{Flowrate, m}^3/\text{s} = (\text{Area, m}^2)(\text{Velocity, m/s}) \quad \text{or} \quad Q = AV \quad \text{where } Q = \text{flowrate, } A = \text{area, } V = \text{velocity}$$

$$\text{Food-to-Microorganism Ratio} = \frac{\text{BOD}_5, \text{ kg/d}}{\text{MLVSS, kg}}$$

$$\text{Force, N} = (\text{Pressure, Pa})(\text{Area, m}^2)$$

$$\text{Hardness, as mg CaCO}_3/\text{L} = \frac{(\text{Titrant Volume, mL})(1000 \text{ mL/L})}{\text{Sample Volume, mL}} \qquad \begin{array}{l}\text{Only when the titration factor} \\ \text{is 1.00 of EDTA}\end{array}$$

$$\text{Horsepower, Brake (bhp)} = \frac{(\text{Flow, gpm})(\text{Head, ft})}{(3960)(\text{Decimal Pump Efficiency})}$$

$$\text{Horsepower, Motor (mhp)} = \frac{(\text{Flow, gpm})(\text{Head, ft})}{(3960)(\text{Decimal Pump Efficiency})(\text{Decimal Motor Efficiency})}$$

$$\text{Horsepower, Water (whp)} = \frac{(\text{Flow, gpm})(\text{Head, ft})}{3960}$$

$$\text{Hydraulic Loading Rate, m}^3/\text{m}^2 \cdot \text{d} = \frac{\text{Total Flow Applied, m}^3/\text{d}}{\text{Area, m}^2}$$

$$\text{Leakage, L/d} = \frac{\text{Volume, L}}{\text{Time, days}}$$

$$\text{Liters/Capita/Day} = \frac{\text{Volume of Water Produced, L/d}}{\text{Population}}$$

$$\text{Mass, kg} = \frac{(\text{Volume, m}^3)(\text{Concentration, mg/L})}{1000}$$

$$\text{Mass Flux, kg/d} = \frac{(\text{Volume, m}^3/\text{d})(\text{Concentration, mg/L})}{1000}$$

$$\text{Mean Cell Residence Time (MCRT) or Solids Retention Time (SRT), days} = \frac{\text{Aeration Tank TSS, kg} + \text{Clarifier TSS, kg}}{\text{TSS Wasted, kg/d} + \text{Effluent TSS, kg/d}}$$

$$\text{Molarity} = \frac{\text{Moles of Solute}}{\text{Liters of Solution}}$$

$$\text{Normality} = \frac{\text{Number of Equivalent Weights of Solute}}{\text{Liters of Solution}}$$

$$\text{Number of Equivalent Weights} = \frac{\text{Total Weight}}{\text{Equivalent Weight}}$$

$$\text{Number of Moles} = \frac{\text{Total Weight}}{\text{Molecular Weight}}$$

$$\text{Organic Loading Rate, kg/m}^3 \cdot \text{d} = \frac{\text{Organic Load, kg BOD}_5/\text{d}}{\text{Volume}}$$

$$\text{Organic Loading Rate} - \text{RBC, kg/m}^2 \cdot \text{d} = \frac{\text{Organic Load, kg BOD}_5/\text{d}}{\text{Surface Area of Media, m}^2}$$

$$\text{Organic Loading Rate-Trickling Filter, kg/m}^3 \cdot \text{d} = \frac{\text{Organic Load, kg BOD}_5/\text{d}}{\text{Volume, m}^3}$$

$$\text{Oxygen Uptake Rate/Oxygen Consumption Rate, mg/L/min} = \frac{\text{Oxygen Usage, mg/L}}{\text{Time, minute}}$$

$$\text{Population Equivalent, Organic} = \frac{(\text{Flow, m}^3/\text{d})(\text{BOD, mg/L})}{(1000)(.077 \text{ kg BOD/d/person})}$$

$$\text{Power, kW} = \frac{(\text{Flow, L/sec})(\text{Head, m})(9.8)}{1000}$$

$$\text{Recirculation Ratio-Tricking Filter} = \frac{\text{Recirculated Flow}}{\text{Primary Effluent Flow}}$$

$$\text{Reduction in Flow, \%} = \frac{(\text{Original Flow} - \text{Reduced Flow})(100\%)}{\text{Original Flow}}$$

$$\text{Reduction of Volatile Solids, \%} = \frac{(\text{In} - \text{Out})(100\%)}{\text{In} - (\text{In} \times \text{Out})}$$ All information (In and Out) must be in decimal form.

$$\text{Removal, \%} = \frac{(\text{In} - \text{Out})(100)}{\text{In}}$$

$$\text{Return Rate, \%} = \frac{(\text{Return Flowrate})(100\%)}{\text{Influent Flowrate}}$$

$$\text{Return Sludge Rate-Solids Balance} = \frac{(\text{MLSS})(\text{Flowrate})}{\text{Return Activated Sludge Suspended Solids} - \text{MLSS}}$$

$$\text{Slope, \%} = \frac{\text{Drop or Rise}}{\text{Distance}} \times 100\%$$

$$\text{Sludge Density Index} = \frac{100}{\text{SVI}}$$

$$\text{Sludge Volume Index, mL/g} = \frac{(\text{SSV}_{30}, \text{ mL/L})(1000 \text{ mg/g})}{\text{MLSS, mg/L}}$$

$$\text{Solids, mg/L} = \frac{(\text{Dry Solids, grams})(1000 \text{ mg/g})(1000 \text{ mL/L})}{\text{Sample Volume, mL}}$$

$$\text{Solids Concentration, mg/L} = \frac{\text{Weight, mg}}{\text{Volume, L}}$$

$$\text{Solids Loading Rate, kg/d/m}^2 = \frac{\text{Solids Applied, kg/d}}{\text{Surface Area, m}^2}$$

Solids Retention Time (SRT): *see* Mean Cell Residence Time (MCRT)

$$\text{Specific Gravity} = \frac{\text{Specific Weight of Substance, kg/L}}{\text{Specific Weight of Water, kg/L}}$$

Specific Oxygen Uptake Rate/Respiration Rate, $(mg/g)/h = \dfrac{OUR, mg/L/min\ (60\ min)}{MLVSS, g/L\ (1\ h)}$

Surface Loading Rate or Surface Overflow Rate, $L/m^2 \cdot d = \dfrac{Flow, L/d}{Area, m^2}$

Three Normal Equation $= (N_1 \times V_1) + (N_2 \times V_2) = (N_3 \times V_3)$, where $V_1 + V_2 = V_3$

Two Normal Equation $= N_1 \times V_1 = N_2 \times V_2$, where $N =$ concentration (normality), $V =$ volume or flow

Velocity, $m/s = \dfrac{Flowrate, m^3/sec}{Area, m^2}$ or $\dfrac{Distance, m}{Time, second}$

Volatile Solids, $\% = \dfrac{(Dry\ solids, g - Fixed\ solids, g)(100)}{Dry\ solids, g}$

Volume of Cone $= (1/3)(0.785)(Diameter^2)(Height)$

Volume of Cylinder $= (0.785)(Diameter^2)(Height)$

Volume of Rectangular Tank $= (Length)(Width)(Height)$

Waste Milliequivalent $= (mL)(Normality)$

Watts (DC circuit) $= (Volts)(Amps)$

Watts (AC circuit) $= (Volts)(Amps)(Power\ Factor)$

Weir Overflow Rate, $L/m \cdot d = \dfrac{Flow, L/d}{Weir\ Length, m}$

Wire-to-Water Efficiency, $\% = \dfrac{Water\ Horsepower, hp}{Power\ Input, hp\ or\ Motor\ hp} \times 100$

Wire-to-Water Efficiency, $\% = \dfrac{(Flow, gpm)(Total\ Dynamic\ Head, ft)(0.746\ kW/hp)}{(3960)(Electrical\ Demand, kW)} \times 100$

Formula/Conversion Table for Wastewater Treatment, Industrial, Collection, and Laboratory Exams in U.S. Customary Units

$$\text{Alkalinity, as mg CaCO}_3/\text{L} = \frac{(\text{Titrant Volume, mL})(\text{Acid Normality})(50\ 000\ \text{mg/equiv})}{\text{Sample Volume, mL}}$$

$$\text{Amps} = \frac{\text{Volts}}{\text{Ohms}}$$

$$*\text{Area of Circle} = (0.785)(\text{Diameter}^2) \text{ or}$$
$$= (\pi)(\text{Radius}^2)$$

$$\text{Area of Cone (lateral area)} = (\pi)(\text{Radius})\sqrt{\text{Radius}^2 + \text{Height}^2}$$

$$\text{Area of Cone (total surface area)} = (\pi)(\text{Radius})(\text{Radius} + \sqrt{\text{Radius}^2 + \text{Height}^2})$$

$$\text{Area of Cylinder (total exterior surface area)} = (\text{Surface Area of End \#1}) + (\text{Surface Area of End \#2}) + [(\pi)(\text{Diameter})(\text{Height or Depth})]$$

$$*\text{Area of Rectangle} = (\text{Length})(\text{Width})$$

$$*\text{Area of Right Triangle} = \frac{(\text{Base})(\text{Height})}{2}$$

$$\text{Average (arithmetic mean)} = \frac{\text{Sum of All Terms}}{\text{Number of Terms}}$$

$$\text{Average (geometric mean)} = [(X_1)(X_2)(X_3)(X_4)(X_n)]^{1/n} \text{ The } n\text{th root of the product of } n \text{ numbers}$$

$$\text{Biochemical Oxygen Demand (unseeded), mg/L} = \frac{[(\text{Initial Dissolved Oxygen, mg/L}) - (\text{Final Dissolved Oxygen, mg/L})][300\ \text{mL}]}{\text{Sample Volume, mL}}$$

$$\text{Chemical Feed Pump Setting, \% Stroke} = \frac{\text{Desired Flow}}{\text{Maximum Flow}} \times 100\%$$

$$\text{Chemical Feed Pump Setting, mL/min} = \frac{(\text{Flow, mgd})(\text{Dose, mg/L})(3.785\ \text{L/gal})(1\ 000\ 000\ \text{gal/mil. gal})}{(\text{Liquid, mg/mL})(24\ \text{hr/d})(60\ \text{min/hr})}$$

*Pie wheel format for this equation is available at the end of this document.

Circumference of Circle $= (\pi)(\text{Diameter})$ or
$$= 2(\pi)(\text{Radius})$$

$$\text{Composite Sample Single Portion} = \frac{(\text{Instantaneous Flow})(\text{Total Sample Volume})}{(\text{Number of Portions})(\text{Average Flow})}$$

$$\text{Cycle Time, minute} = \frac{\text{Storage Volume, gal}}{\text{Pump Capacity, gpm} - \text{Wet Well Inflow, gpm}}$$

Degrees Celsius $= (\text{Degrees Fahrenheit} - 32)(\tfrac{5}{9})$
$$= \frac{(°F - 32)}{1.8}$$

Degrees Fahrenheit $= (\text{Degrees Celsius})(\tfrac{9}{5}) + 32$
$$= (\text{Degrees Celsius})(1.8) + 32$$

$$\text{Detention Time} = \frac{\text{Volume}}{\text{Flow}} \quad \text{Units must be compatible.}$$

Dose = Demand + Residual

*Electromotive force (EMF), $V = (\text{Current, amps})(\text{Resistance, ohms})$ or $E = IR$

$$*\text{Feed Rate, lb/d} = \frac{(\text{Dosage, mg/L})(\text{Capacity, mgd})(8.34 \text{ lb/mil. gal})}{\text{Purity, \% expressed as a decimal}}$$

$$\text{Filter Backwash Rise Rate, in./min} = \frac{(\text{Backwash Rate, gpm/sq ft})(12 \text{ in./ft})}{7.48 \text{ gal/cu ft}}$$

$$\text{Filter Flowrate or Backwash Rate, gpm/sq ft} = \frac{\text{Flow, gpm}}{\text{Filter Area, sq ft}}$$

$$\text{Filter Yield, lb/hr/sq ft} = \frac{(\text{Solids Loading, lb/d})(\text{Recovery, \% expressed as a decimal})}{(\text{Filter Operation, hr/d})(\text{Area, sq ft})}$$

*Flowrate, cfs $= (\text{Area, sq ft})(\text{Velocity, ft/sec})$ or $Q = AV$ Units must be compatible.

$$\text{Food-to-Microorganism Ratio} = \frac{\text{BOD}_5, \text{ lb/d}}{\text{MLVSS, lb}}$$

*Force, lb $= (\text{Pressure, psi})(\text{Area, sq in.})$

*Pie wheel format for this equation is available at the end of this document.

$$\text{Gallons/Capita/Day} = \frac{\text{Volume of Water Produced, gpd}}{\text{Population}}$$

$$\text{Hardness, as mg CaCO}_3/\text{L} = \frac{(\text{Titrant Volume, mL})(1000 \text{ mL/L})}{\text{Sample Volume, mL}} \quad \text{Only when the titration factor is 1.00 of EDTA}$$

$$\text{Horsepower, Brake (bhp)} = \frac{(\text{Flow, gpm})(\text{Head, ft})}{(3960)(\text{Pump Efficiency, \% expressed as a decimal})}$$

$$\text{Horsepower, Motor (mhp)} = \frac{(\text{Flow, gpm})(\text{Head, ft})}{(3960)(\text{Pump Efficiency, \% expressed as a decimal})(\text{Motor Efficiency, \% expressed as a decimal})}$$

$$*\text{Horsepower, Water (whp)} = \frac{(\text{Flow, gpm})(\text{Head, ft})}{3960}$$

$$\text{Hydraulic Loading Rate, gpd/sq ft} = \frac{\text{Total Flow Applied, gpd}}{\text{Area, sq ft}}$$

$$\text{Leakage, gpd} = \frac{\text{Volume, gal}}{\text{Time, d}}$$

$$*\text{Mass, lb} = (\text{Volume, mil. gal})(\text{Concentration, mg/L})(8.34 \text{ lb/mil. gal})$$

$$*\text{Mass Flux, lb/d} = (\text{Flow, mgd})(\text{Concentration, mg/L})(8.34 \text{ lb/mil. gal})$$

$$\begin{matrix}\text{Mean Cell Residence Time (MCRT) or} \\ \text{Solids Retention Time (SRT), days}\end{matrix} = \frac{\text{Aeration Tank TSS, lb} + \text{Clarifier TSS, lb}}{\text{TSS Wasted, lb/d} + \text{Effluent TSS, lb/d}}$$

$$\text{Milliequivalent} = (\text{mL})(\text{Normality})$$

$$\text{Molarity} = \frac{\text{Moles of Solute}}{\text{Liters of Solution}}$$

$$\text{Motor Efficiency, \%} = \frac{\text{Brake hp}}{\text{Motor hp}} \times 100\%$$

$$\text{Normality} = \frac{\text{Number of Equivalent Weights of Solute}}{\text{Liters of Solution}}$$

*Pie wheel format for this equation is available at the end of this document.

$$\text{Number of Equivalent Weights} = \frac{\text{Total Weight}}{\text{Equivalent Weight}}$$

$$\text{Number of Moles} = \frac{\text{Total Weight}}{\text{Molecular Weight}}$$

$$\text{Organic Loading Rate, lb BOD}_5 \text{/d/cu ft} = \frac{\text{Organic Load, lb BOD}_5 \text{/d}}{\text{Volume, cu ft}}$$

$$\text{Organic Loading Rate-RBC, lb BOD}_5 \text{/d/1000 sq ft} = \frac{\text{Organic Load, lb BOD}_5 \text{/d}}{\text{Surface Area of Media, 1000 sq ft}}$$

$$\text{Organic Loading Rate-Tricking Filter, lb BOD}_5 \text{/d/1000 cu ft} = \frac{\text{Organic Load, lb BOD}_5 \text{/d}}{\text{Volume, 1000 cu ft}}$$

$$\text{Oxygen Uptake Rate or Oxygen Consumption Rate, mg/L/min} = \frac{\text{Oxygen Usage, mg/L}}{\text{Time, min}}$$

$$\text{Population Equivalent, Organic} = \frac{(\text{Flow, mgd})(\text{BOD, mg/L})(8.34 \text{ lb/mil. gal})}{\text{BOD/d/person, lb}}$$

$$\text{Recirculation Ratio-Tricking Filter} = \frac{\text{Recirculated Flow}}{\text{Primary Effluent Flow}}$$

$$\text{Reduction in Flow, \%} = \left(\frac{\text{Original Flow} - \text{Reduced Flow}}{\text{Original Flow}} \right) \times 100\%$$

$$\text{Reduction of Volatile Solids, \%} = \left(\frac{\text{In} - \text{Out}}{\text{In} - (\text{In} \times \text{Out})} \right) \times 100\% \qquad \text{All information (In and Out) must be in decimal form.}$$

$$\text{Removal, \%} = \left(\frac{\text{In} - \text{Out}}{\text{In}} \right) \times 100\%$$

$$\text{Return Rate, \%} = \frac{\text{Return Flowrate}}{\text{Influent Flowrate}} \times 100\%$$

$$\text{Return Sludge Rate-Solids Balance} = \frac{(\text{MLSS})(\text{Flowrate})}{\text{Return Activated Sludge Suspended Solids} - \text{MLSS}}$$

$$\text{Slope, \%} = \frac{\text{Drop or Rise}}{\text{Distance}} \times 100\%$$

$$\text{Sludge Density Index} = \frac{100}{\text{SVI}}$$

$$\text{Sludge Volume Index (SVI), mL/g} = \frac{(\text{SSV}_{30}, \text{mL/L})(1000 \text{ mg/g})}{\text{MLSS, mg/L}}$$

$$\text{Solids, mg/L} = \frac{(\text{Dry Solids, g})(1000 \text{ mg/g})(1000 \text{ mL/L})}{\text{Sample Volume, mL}}$$

$$\text{Solids Concentration, mg/L} = \frac{\text{Weight, mg}}{\text{Volume, L}}$$

$$\text{Solids Loading Rate, lb/d/sq ft} = \frac{\text{Solids Applied, lb/d}}{\text{Surface Area, sq ft}}$$

Solids Retention Time (SRT): *see* Mean Cell Residence Time (MCRT)

$$\text{Specific Gravity} = \frac{\text{Specific Weight of Substance, lb/gal}}{\text{Specific Weight of Water, lb/gal}}$$

$$\text{Specific Oxygen Uptake Rate or Respiration Rate, (mg/g)/h} = \frac{\text{OUR, mg/L/min (60 min)}}{\text{MLVSS, g/L (1 hr)}}$$

$$\text{Surface Loading Rate or Surface Overflow Rate, gpd/sq ft} = \frac{\text{Flow, gpd}}{\text{Area, sq ft}}$$

$$\text{Three Normal Equation} = (N_1 \times V_1) + (N_2 \times V_2) = (N_3 \times V_3) \quad \text{where } V_1 + V_2 = V_3$$

$$\text{Two Normal Equation} = N_1 \times V_1 = N_2 \times V_2 \quad \text{where } N = \text{normality}, V = \text{volume or flow}$$

$$\text{Velocity, ft/sec} = \frac{\text{Flowrate, cu ft/sec}}{\text{Area, sq ft}} \quad \text{or} \quad \frac{\text{Distance, ft}}{\text{Time, sec}}$$

$$\text{Volatile Solids, \%} = \left(\frac{\text{Dry Solids, g} - \text{Fixed Solids, g}}{\text{Dry Solids, g}} \right) \times 100\%$$

*Pie wheel format for this equation is available at the end of this document.

*Volume of Cone $= (1/3)(0.785)(\text{Diameter}^2)(\text{Height})$ or
$$= (1/3)[(\pi)(\text{Radius}^2)(\text{Height})]$$

*Volume of Cylinder $= (0.785)(\text{Diameter}^2)(\text{Height})$ or
$$= (\pi)(\text{Radius}^2)(\text{Height})$$

*Volume of Rectangular Tank $= (\text{Length})(\text{Width})(\text{Height})$

Watts (AC circuit) $= (\text{Volts})(\text{Amps})(\text{Power Factor})$

Watts (DC circuit) $= (\text{Volts})(\text{Amps})$

$$\text{Weir Overflow Rate, gpd/ft} = \frac{\text{Flow, gpd}}{\text{Weir Length, ft}}$$

$$\text{Wire-to-Water Efficiency, \%} = \frac{\text{Water Horsepower, hp}}{\text{Power Input, hp or Motor hp}} \times 100\%$$

$$\text{Wire-to-Water Efficiency, \%} = \frac{(\text{Flow, gpm})(\text{Total Dynamic Head, ft})(0.746 \text{ kW/hp})}{(3960)(\text{Electrical Demand, kW})} \times 100\%$$

*Pie wheel format for this equation is available at the end of this document.

Conversion Factors

1 ac	=	4046.9 m^2 or 43 560 sq ft	1 lb/d/cu ft	=	16.02 kg/m^3·d
1 ac ft	=	326 000 gal	1 lb/d/cu ft	=	40.74 L/m^2·d
1 cfs	=	0.028 m^3/s	1 lb/d/sq ft	=	4.882 kg/m^2·d
1 cu ft	=	7.48 gal	1 lb/gal	=	0.120 kg/L
	=	62.4 lb	1 lb/hr/sq ft	=	4.882 kg/m^2·h
1 cu yd	=	27 cu ft	1 m head	=	9.8 kPa
1 cfs	=	0.646 mgd	1 m^2	=	1.19 sq yd
1 ft	=	0.305 m	1 m^3	=	1000 kg
1 ft H$_2$O	=	0.433 psi		=	1000 L
1 ft/sec	=	0.305 m/s		=	219.97 gal
1 gal	=	3.79 L	1 m^3/s	=	19.01 mgd
	=	8.34 lb*	1 metric ton	=	2204.62 lb
1 gpd	=	0.004 m^3/d	1 mile	=	5280 ft
1 gpd/ft	=	0.012 m^3/m·d	1 mgd	=	694 gpm
1 gpd/sq ft	=	0.041 m^3/m^2·d or m/d		=	1.55 cfs
1 gpm	=	0.063 L/s		=	3.785 ML/d
1 gpm/sq ft	=	0.679 L/m^2·s	1 mil. gal	=	3.785 ML/d
	=	58.67 m^3/m^2·d	PE, hydraulic	=	378.5 L/person/d
1 gr/gal	=	17.1 mg/L	PE, organic	=	0.077 kg/BOD/person/d
1 ha	=	10 000 m^2	1 psi	=	2.31 ft of water
1 hp	=	0.746 kW		=	6.89 kPa
	=	746 W	1 sq ft	=	0.093 m^2
	=	33 000 ft lb/min	1 sq in.	=	645.2 mm^2
1 in.	=	25.4 mm	1 ton	=	2000 lb
1 lb	=	0.454 kg	1%	=	10 000 mg/L
1 lb/d	=	0.454 kg/d	π or pi	=	3.14159

*For water and other fluids with the same density as water.

*Pie Wheels:

- To find the quantity above the horizontal line: multiply the pie wedges below the line together.
- To solve for one of the pie wedges below the horizontal line: cover that pie wedge, then divide the remaining pie wedge(s) into the quantity above the horizontal line.

Given units must match the units shown in the pie wheel.

Area of Circle

Area of Rectangle

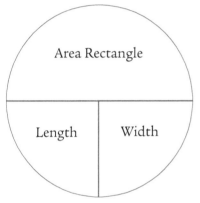

Area of Right Triangle

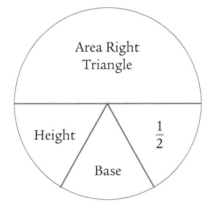

Electromotive Force (EMF), Volts

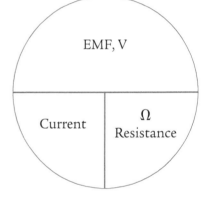

Feed Rate, lb/d

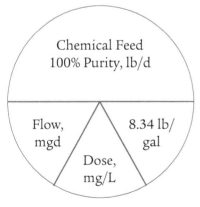

Flowrate, cfs

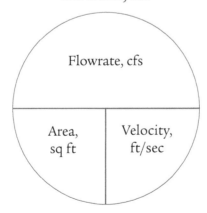

Force, lb

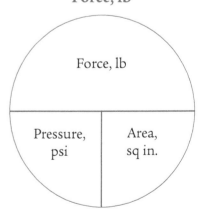

Horsepower, water

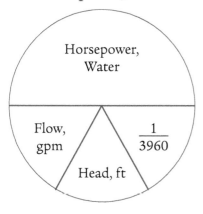

Mass, lb

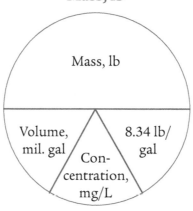

Mass Flux, lb/d

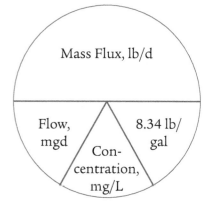

Volume of Cone

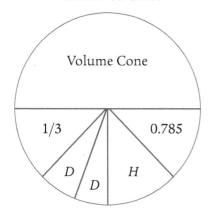

Volume of Cylinder

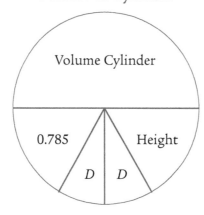

Volume of Rectangular Tank

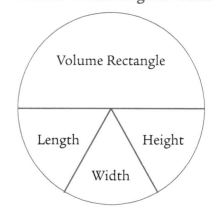

NEED-TO-KNOW CRITERIA AND SAMPLE QUESTIONS

Security, Safety, and Administrative Procedures

Class I Job Tasks

- Adhere to established safety procedures (e.g., lockout/tagout, confined spaces, hazard communication, fall protection),
- Complete operation reports,
- Complete required regulatory reports,
- Conduct routine security checks,
- Ensure compliance with all applicable regulations,
- Generate maintenance reports (e.g., daily, monthly, annual),
- Implement spill notification policy,
- Inspect SCBA equipment,
- Receive chemical deliveries and store,
- Schedule routine activities (e.g., maintenance, operations), and
- Update standard operating procedures (SOPs).

Class I Types of Knowledge Required to Perform Job Tasks

Types of Knowledge	Level of Knowledge
Bacteriological laboratory testing (e.g., coliform, fecal, *E coli*)	Basic
Biological laboratory testing (e.g., BOD, SOUR, CBOD)	Basic
Chemical handling and storage	Basic
Chemical laboratory testing (e.g., ammonia, phosphorus, alkalinity)	Basic
Chlorinators (e.g., gas, liquid)	Basic
Clarifiers	Basic
Comminuters	Basic
Conveyors	Basic
Dewatering equipment (e.g., centrifuges, presses, drying beds)	Basic
Documentation and recordkeeping	Basic
Effluent disposal and monitoring requirements	Basic

Types of Knowledge	Level of Knowledge
Electrical principles (e.g., troubleshooting breakers, relays, circuits)	Basic
Emergency preparedness	Basic
Grit removal processes (e.g., grit chamber, aeration, cyclone)	Basic
Heavy equipment (e.g., operation, preventive maintenance)	Basic
Industry safety practices (e.g., PPE, confined spaces, fall arrest, lockout/tagout)	Basic
Influent monitoring and waste characteristics	Basic
Maintenance practices (e.g., preventive, reactive, predictive)	Basic
Ozone generation equipment	Basic
Physical laboratory testing (e.g., temperature, solids, dissolved oxygen)	Basic
Pneumatic principles (e.g., troubleshooting actuators, compressors, sprayers)	Basic
Primary treatment processes (e.g., ponds, sedimentation basins)	Basic
Principles of asset management (e.g., preventive, reactive, predictive maintenance)	Basic
Quality control/quality assurance practices	Basic
Screening technology (e.g., bar screens, microscreens)	Basic
Secondary treatment processes (e.g., activated sludge, MBR, SBR)	Basic
Tertiary treatment processes (e.g., media filtration, disinfection, post-aeration, reclaimed recharge)	Basic
Treatment equipment (e.g., pumps, motors, generators)	Basic

Class I Sample Questions

1. What is the definition of *hazard*?

   a) The capacity to cause harm
   b) The change or probability that a person will experience harm
   c) The practical assurance of no harm
   d) Risk/safety

2. What is the primary purpose of a safety data sheet (SDS)?

   a) To assist with the maintenance of proper inventories of materials
   b) To provide a record of the proper storage of materials

c) To provide health and safety information regarding the use and handling of specific materials

d) To identify when materials have expired and need to be replaced

3. A sewer system shall not be entered without a self-contained breathing apparatus (SCBA) if the hydrogen sulfide (H_2S) level exceeds

a) 10 ppm
b) 100 ppm
c) 200 ppm
d) 500 ppm

4. Nobody may enter a confined space, as defined by the Occupational Safety and Health Administration or the employer, unless the proposed entrant has first undergone training in

a) The wearing, use, and maintenance of enhanced level B personal protective equipment (PPE).
b) The details of the facility confined space entry procedures.
c) Emergency first aid related to the contaminants expected to be present.
d) Maintenance of the appropriate rescue equipment.

5. What type of kit may be used to repair a chlorine leak from a 1-ton container?

a) Type A
b) Type B
c) Type C
d) Type D

6. The exhaust vents for a chlorine or sulfur dioxide cylinder room are located

a) As close to the chlorine or sulfur dioxide cylinders as possible.
b) Near the cylinder bonnets about two-thirds of the way up the wall.
c) Near the floor.
d) Immediately adjacent to entrances.

7. Which class of extinguisher should be used for electrical fires?

a) Class A
b) Class B
c) Class C
d) Class D

Answers

1. **Answer:** A

    Hazard is the capacity to cause harm. Risk is the chance or probability that a person will experience harm.

    Reference: Water Environmental Federation (2013) *Safety, Health, and Security in Wastewater Systems,* 6th ed.; Manual of Practice No. 1; Water Environment Federation: Alexandria, Virginia; p 37.

2. **Answer:** C

    The safety data sheet provides the properties of each chemical; the physical, health, and environmental health hazards; protective measures; and safety precautions for handling, storing, and transporting the chemical.

    The term material safety data sheet (MSDS) is for chemical data sheets written before the Global Harmonization Standard (GHS) in 2012. The term SDS is for chemical data sheets written after September 2012 that are in compliance with the new 16 section form of the GHS. The Occupational Safety and Health Administration uses the term safety data sheet in the current HAZCOM standard that is referenced in the answer.

    Reference: 29 CFR 1910.1200(g).

3. **Answer:** A

    Entrants should be trained in the use of, and be equipped with, atmospheric monitoring equipment that sounds an audible alarm, in addition to its visual readout, whenever one of the following conditions are encountered: oxygen concentration less than 19.5%; flammable gas or vapor at 10% or more of the lower flammable limit (LFL); or hydrogen sulfide or carbon monoxide at or above 10 ppm or 35 ppm, respectively, measured as an 8-hour time-weighted average.

    Reference: 29 CFR 1910.146, Appendix E.

4. **Answer:** B

    No one may enter a confined space unless properly trained as an entrant. In addition, he/she must follow confined space entry procedures, including ventilation, gas monitoring, lockout/tagout procedures, and using the proper level of PPE and rescue equipment.

    Reference: Water Environment Federation (2013) *Safety, Health, and Security in Wastewater Systems,* 6th ed.; Manual of Practice No. 1; Water Environment Federation: Alexandria, Virginia; p 72.

5. **Answer:** B

A type A kit contains tools and materials to repair a 150-lb cylinder. A type B kit contains tools and materials for a 1-ton container. A type C kit contains tools and materials to repair leaks from a rail car. There is no such thing as a type D kit.

Reference: Water Environment Federation (2013) *Safety, Health, and Security in Wastewater Systems*, 6th ed.; Manual of Practice No. 1; Water Environment Federation: Alexandria, Virginia; p 197.

6. **Answer:** C

Chlorine gas and sulfur dioxide gas are heavier than air. If a leak occurs, these gases will tend to accumulate near the floor. For this reason, exhaust vents are placed near the floor. Gas rooms constructed before 1986 may or may not have a chlorine scrubber to treat chlorine before releasing vented air to the atmosphere. After 1986, scrubbers were required under National Fire Protection Association (NFPA) standards.

Reference: Water Environment Federation (2008) *Operation of Municipal Wastewater Treatment Plants*, 6th ed.; Manual of Practice No. 11; Water Environment Federation: Alexandria, Virginia; p 5-14.

7. **Answer:** C

The following are the four classes of fires:

  Class A—ordinary combustibles
  Class B—combustible liquids and vapors
  Class C—electrical fires
  Class D—combustible metals

Reference: Water Environment Federation (2008) *Operation of Municipal Wastewater Treatment Plants*, 6th ed.; Manual of Practice No. 11; Water Environment Federation: Alexandria, Virginia; p 5-44.

Class II Job Tasks

- Adhere to established safety procedures (e.g., lockout/tagout, confined spaces, hazard communication, fall protection),
- Complete operation reports,
- Complete required regulatory reports,
- Conduct routine security checks,
- Ensure compliance with all applicable regulations,
- Generate maintenance reports (e.g., daily, monthly, annual),
- Implement spill notification policy,

- Inspect SCBA equipment,
- Receive chemical deliveries and store,
- Schedule routine activities (e.g., maintenance, operations), and
- Update standard operating procedures (SOPs).

Class II Types of Knowledge Required to Perform Job Tasks

Types of Knowledge	Level of Knowledge
Aeration principles (e.g., mixing, mechanical, diffusers)	Basic
Bacteriological laboratory testing (e.g., coliform, fecal, *E coli*)	Basic
Biological laboratory testing (e.g., BOD, SOUR, CBOD)	Basic
Biosolids disposal and monitoring requirements	Basic
Chemical handling and storage	Basic
Chemical laboratory testing (e.g., ammonia, phosphorus, alkalinity)	Basic
Chlorinators (e.g., gas, liquid)	Basic
Clarifiers	Basic
Comminuters	Basic
Conveyors	Basic
Dewatering equipment (e.g., centrifuges, presses, drying beds)	Basic
Documentation and recordkeeping	Basic
Effluent disposal and monitoring requirements	Basic
Electrical principles (e.g., troubleshooting breakers, relays, circuits)	Basic
Emergency preparedness	Basic
Grit removal processes (e.g., grit chamber, aeration, cyclone)	Basic
Heavy equipment (e.g., operation, preventive maintenance)	Basic
Industry safety practices (e.g., PPE, confined spaces, fall arrest, lockout/tagout)	Basic
Influent monitoring and waste characteristics	Basic
Maintenance practices (e.g., preventive, reactive, predictive)	Basic
Ozone generation equipment	Basic
Physical laboratory testing (e.g., temperature, solids, dissolved oxygen)	Basic
Pneumatic principles (e.g., troubleshooting actuators, compressors, sprayers)	Basic
Primary treatment processes (e.g., ponds, sedimentation basins)	Basic
Principles of asset management (e.g., preventive, reactive, predictive maintenance)	Basic
Quality control/quality assurance practices	Basic

Types of Knowledge	Level of Knowledge
Screening technology (e.g., bar screens, microscreens)	Basic
Secondary treatment processes (e.g., activated sludge, MBR, SBR)	Basic
Tertiary treatment processes (e.g., media filtration, disinfection, post-aeration, reclaimed recharge)	Basic
Treatment equipment (e.g., pumps, motors, generators)	Basic

Class II Sample Questions

1. What personal protective equipment (PPE) does the Chlorine Institute recommend for emergency response to a chlorine release when liquid chlorine is not involved?

    a) Level B
    b) Enhanced Level B
    c) Level C
    d) Enhanced Level C

2. The Occupational Safety and Health Administration defines a "confined space" using three conditional criteria. Which of the following correctly identifies those three criteria?

    a) The space exceeds 25 cu ft (0.75 m³) in size with a depth of at least 4 ft (13.2 m); the space has fewer than two adequate means of egress; the space has no natural ventilation.
    b) The space is deeper than 5 ft (16.5 m) and has no side means of access or egress; the space is not adequately ventilated; the space was not designed for human occupancy under any circumstances.
    c) The space is deeper than 5 ft (15.5 m) and has a volume exceeding 25 cu ft (0.75 m³); the space does not contain adequate means of access or egress; the space has no adequate means of providing rescue in case of emergency.
    d) The space must be of adequate size and have a configuration where a worker can enter the space; the space has a limited means of access or egress; the space was not designed for continuous human occupancy.

3. Which of the following is considered to be MOST important for preventing personal exposure to biological hazards in a water resource recovery facility?

    a) Proper disinfection of equipment after use
    b) Proper disinfection of collected samples
    c) Proper personal protective equipment and proper hygiene
    d) Proper prophylactic use of antibiotics

4. At what concentration in air is chlorine gas likely to cause toxic pneumonitis and pulmonary edema?

    a) Less than 40 ppm
    b) More than 40 ppm, but less than 60 ppm
    c) More than 60 ppm, but less than 80 ppm
    d) More than 80 ppm

Answers

1. **Answer:** A

    The Chlorine Institute recommends Level B PPE for chlorine releases that do not involve liquid chlorine and enhanced Level B PPE for those releases that do involve liquid chlorine.

    Reference: Water Environment Federation (2013) *Safety, Health, and Security in Wastewater Systems*, 6th ed.; Manual of Practice No. 1; Water Environment Federation: Alexandria, Virginia; p 70.

2. **Answer:** D

    The Occupational Safety and Health Administration defines a "confined space" using three conditional criteria. First, the space must be of adequate size and have a configuration where a worker can enter the space. Second, the space has a limited means of access or egress. Third, the space was not designed for continuous human occupancy.

    Reference: Water Environment Federation (2013) *Safety, Health, and Security in Wastewater Systems*, 6th ed.; Manual of Practice No. 1; Water Environment Federation: Alexandria, Virginia; p 172.

3. **Answer:** C

    Because of their daily exposure to wastewater-contaminant environments, water resource recovery facility personnel have a higher incidence of potential exposure to pathogens than the general public. For most workers, however, the risk of developing a disease is relatively low. Proper personal hygiene is still critical because infections may occur without symptoms and antibodies to bacteria and viruses may develop without illness symptoms being readily apparent.

    Reference: Water Environment Federation (2013) *Safety, Health, and Security in Wastewater Systems*, 6th ed.; Manual of Practice No. 1; Water Environment Federation: Alexandria, Virginia; p 257.

4. **Answer:** B

Chlorine gas will cause acute respiratory hazards from inhalation. More than 40 ppm but less than 60 ppm of chlorine gas causes toxic pneumonitis and pulmonary edema. At 430 ppm, an exposure is lethal within 30 minutes and at 1000 ppm, it is fatal within minutes.

Reference: Water Environment Federation (2013) *Safety, Health, and Security in Wastewater Systems*, 6th ed.; Manual of Practice No. 1; Water Environment Federation: Alexandria, Virginia; pp 69–70.

Class III Job Tasks

- Adhere to established safety procedures (e.g., lockout/tagout, confined spaces, hazard communication, fall protection);
- Assist in the selection of equipment for use in wastewater processing;
- Complete operation reports;
- Complete required regulatory reports;
- Conduct routine security checks;
- Ensure compliance with all applicable regulations;
- Generate maintenance reports (e.g., daily, monthly, annual);
- Implement spill notification policy;
- Inspect SCBA equipment;
- Manage facility staff;
- Participate in studies related to increasing capacity, changes in treatment requirements, or facility upgrades;
- Receive chemical deliveries and store;
- Respond to customer service requests and complaints;
- Schedule routine activities (e.g., maintenance, operations);
- Update spill notification policy; and
- Update standard operating procedures (SOPs).

Class III Types of Knowledge Required to Perform Job Tasks

Types of Knowledge	Level of Knowledge
Aeration principles (e.g., mixing, mechanical, diffusers)	Basic
Bacteriological laboratory testing (e.g., coliform, fecal, *E coli*)	Intermediate
Biological laboratory testing (e.g., BOD, SOUR, CBOD)	Intermediate
Biosolids disposal and monitoring requirements	Intermediate
Chemical handling and storage	Basic

Types of Knowledge	Level of Knowledge
Chemical laboratory testing (e.g., ammonia, phosphorus, alkalinity)	Intermediate
Chlorinators (e.g., gas, liquid)	Intermediate
Clarifiers	Basic
Comminuters	Basic
Conveyors	Basic
Dewatering equipment (e.g., centrifuges, presses, drying beds)	Basic
Documentation and recordkeeping	Intermediate
Effluent disposal and monitoring requirements	Intermediate
Electrical principles (e.g., troubleshooting breakers, relays, circuits)	Basic
Emergency preparedness	Intermediate
Grit removal processes (e.g., grit chamber, aeration, cyclone)	Intermediate
Heavy equipment (e.g., operation, preventive maintenance)	Basic
Industry safety practices (e.g., PPE, confined spaces, fall arrest, lockout/tagout)	Intermediate
Influent monitoring and waste characteristics	Intermediate
Maintenance practices (e.g., preventive, reactive, predictive)	Intermediate
Ozone generation equipment	Basic
Physical laboratory testing (e.g., temperature, solids, dissolved oxygen)	Intermediate
Pneumatic principles (e.g., troubleshooting actuators, compressors, sprayers)	Basic
Primary treatment processes (e.g., ponds, sedimentation basins)	Intermediate
Principles of asset management (e.g., preventive, reactive, predictive maintenance)	Intermediate
Quality control/quality assurance practices	Basic
Screening technology (e.g., bar screens, microscreens)	Basic
Secondary treatment processes (e.g., activated sludge, MBR, SBR)	Intermediate
Tertiary treatment processes (e.g., media filtration, disinfection, post-aeration, reclaimed recharge)	Intermediate
Treatment equipment (e.g., pumps, motors, generators)	Basic

Class III Sample Questions

1. Which of the following is NOT necessarily a component of an effective safety program?

    a) Regular safety briefings
    b) Daily practice
    c) A strong advocate for safety
    d) A monetary reward system

2. Which of the following are the most effective sources of information for managers regarding patterns of ergonomic-related injuries and illnesses when evaluating the success of a safety program?

    a) Employee illness records, employee injury records, and worker's compensation claims data

    b) Employee absenteeism records, employee health care costs data, and average worker's compensation claim costs

    c) Employee absences, employee discipline records, and accident-free-days data

    d) Employee complaints data, supervisor complaints data, and worker's ergonomic equipment request data

Answers

1. **Answer: D**

    An effective safety program has to be practiced daily and must be foremost in a worker's thoughts for the day. An effective safety program includes daily practice, regular safety briefings, and a strong central leader. The central leader can be the head of a group or the safety director of the entire operation at an organizational level or even the supervisor of a work party.

    Reference: Water Environment Federation (2013) *Safety, Health, and Security in Wastewater Systems*, 6th ed.; Manual of Practice No. 1; Water Environment Federation: Alexandria, Virginia; p 131.

2. **Answer: A**

    There are several methods used to identify jobs that are most likely to result in ergonomic disorders. Managers and supervisors should monitor injury and illness records and worker's compensation data to identify patterns of ergonomic-related injuries and illnesses.

    Reference: Water Environment Federation (2013) *Safety, Health, and Security in Wastewater Systems*, 6th ed.; Manual of Practice No. 1; Water Environment Federation: Alexandria, Virginia; p 137.

Class IV Job Tasks

*   Adhere to established safety procedures (e.g., lockout/tagout, confined spaces, hazard communication, fall protection);
*   Assist in the selection of equipment for use in wastewater processing;
*   Assist with budget preparation;
*   Assist with the industrial pretreatment program in regard to effluent quality standards;
*   Complete operation reports;

- Complete required regulatory reports;
- Ensure compliance with all applicable regulations;
- Generate maintenance reports (e.g., daily, monthly, annual);
- Identify personnel training needs;
- Implement spill notification policy;
- Manage facility staff;
- Participate in studies related to increasing capacity, changes in treatment requirements, or facility upgrades;
- Receive chemical deliveries and store;
- Respond to customer service requests and complaints;
- Schedule routine activities (e.g., maintenance, operations);
- Update spill notification policy; and
- Update standard operating procedures (SOPs).

Class IV Types of Knowledge Required to Perform Job Tasks

Types of Knowledge	Level of Knowledge
Aeration principles (e.g., mixing, mechanical, diffusers)	Basic
Bacteriological laboratory testing (e.g., coliform, fecal, *E coli*)	Intermediate
Biological laboratory testing (e.g., BOD, SOUR, CBOD)	Intermediate
Biosolids disposal and monitoring requirements	Advanced
Chemical handling and storage	Basic
Chemical laboratory testing (e.g., ammonia, phosphorus, alkalinity)	Intermediate
Chlorinators (e.g., gas, liquid)	Intermediate
Clarifiers	Basic
Comminuters	Basic
Conveyors	Basic
Dewatering equipment (e.g., centrifuges, presses, drying beds)	Basic
Documentation and recordkeeping	Advanced
Effluent disposal and monitoring requirements	Intermediate
Electrical principles (e.g., troubleshooting breakers, relays, circuits)	Basic
Emergency preparedness	Advanced
Grit removal processes (e.g., grit chamber, aeration, cyclone)	Intermediate
Heavy equipment (e.g., operation, preventive maintenance)	Basic
Industry safety practices (e.g., PPE, confined spaces, fall arrest, lockout/tagout)	Intermediate
Influent monitoring and waste characteristics	Advanced

Types of Knowledge	Level of Knowledge
Maintenance practices (e.g., preventive, reactive, predictive)	Intermediate
Ozone generation equipment	Basic
Physical laboratory testing (e.g., temperature, solids, dissolved oxygen)	Intermediate
Pneumatic principles (e.g., troubleshooting actuators, compressors, sprayers)	Basic
Primary treatment processes (e.g., ponds, sedimentation basins)	Intermediate
Principles of asset management (e.g., preventive, reactive, predictive maintenance)	Intermediate
Quality control/quality assurance practices	Intermediate
Screening technology (e.g., bar screens, microscreens)	Basic
Secondary treatment processes (e.g., activated sludge, MBR, SBR)	Intermediate
Tertiary treatment processes (e.g., media filtration, disinfection, post-aeration, reclaimed recharge)	Intermediate
Treatment equipment (e.g., pumps, motors, generators)	Basic

Class IV Sample Questions

1. Anaerobic digestion gas handling systems are kept under positive pressure for what reason?

    a) To prevent air from entering the system and mixing with digester gas, which could cause an explosion
    b) To minimize leakage of greenhouse gases
    c) To reduce the potential for worker exposure to methane, which can be fatal in low doses
    d) To ensure compliance with the facility's Clean Air Act permit

2. In addition to the mechanical hazards of working around machinery, which of the following environmental hazards are also likely to be faced by maintenance workers at a water resource recovery facility?

    a) Safety device malfunction and heat stress
    b) Toxic substances and radiation
    c) Noise, heat illness, and stress
    d) Temperature extremes and electrical risk

3. Which of the following lists BEST describes the symptoms of heat stroke?

    a) Confusion, loss of consciousness, and seizures
    b) High temperature, strenuous previous activity, and heavy sweating

c) Heavy sweating, cold chills in the victim, and abnormal head pain

d) Headache, nausea, and dizziness

Answers

1. **Answer:** A

   The explosive ratio of air to anaerobic digester gas ranges from approximately 20:1 to 5:1, so avoid mixing digester gas with air at all times. Anaerobic digester gas typically contains approximately 70% methane and 30% carbon dioxide plus nitrogen, hydrogen, hydrogen sulfide, and oxygen.

   Reference: Water Environment Federation (2013) *Safety, Health, and Security in Wastewater Systems*, 6th ed.; Manual of Practice No. 1; Water Environment Federation: Alexandria, Virginia; pp 113–114.

2. **Answer:** C

   In addition to mechanical hazards of working around machinery, there are other hazards that need to be considered. Maintenance personnel must be trained and properly protected from the following hazards as well: noise, heat insulation (asbestos), and heat illness and stress.

   Reference: Water Environment Federation (2013) *Safety, Health, and Security in Wastewater Systems*, 6th ed.; Manual of Practice No. 1; Water Environment Federation: Alexandria, Virginia; pp 147–149.

3. **Answer:** A

   Heat stroke is the most serious heat-related health problem. Heat stroke occurs when the body temperature-regulating system fails and the body core temperature rises to critical levels. This is a medical emergency that may result in death. Signs of heat stroke are confusion, loss of consciousness, and seizures. Workers experiencing heat stroke have a very high body temperature and may stop sweating.

   Reference: Water Environment Federation (2013) *Safety, Health, and Security in Wastewater Systems*, 6th ed.; Manual of Practice No. 1; Water Environment Federation: Alexandria, Virginia; p 150.

Equipment Evaluation and Maintenance

Class I Job Tasks

- Calibrate meters (e.g., flow, pressure sensors);
- Follow safety rules and guidelines when working with chemical equipment;
- Follow safety rules and guidelines when working with mechanical equipment;
- Monitor flowmeters;
- Monitor telemetry systems;
- Perform basic electrical troubleshooting;
- Perform preventive maintenance on equipment;
- Inspect the following equipment:
  - Aeration basins,
  - Aeration systems (e.g., blowers, surface aerators, diffusers),
  - Aerobic digesters,
  - Air compressors,
  - Analyzers (e.g., dissolved oxygen, pH, H_2S, ORP),
  - Bar screens,
  - Chemical feed systems (e.g., polymer, ferric),
  - Chlorination systems,
  - Clarifiers/sedimentation basins,
  - Gates and valves,
  - Generators,
  - Grit removal processes,
  - Hand tools,
  - Heavy equipment,
  - Hoists and cranes,
  - Instrumentation (e.g., flow, pressure, telemetry),
  - Mixers,
  - Motors,
  - Ponds/lagoons,
  - Power tools,
  - Pumps—centrifugal,
  - Pumps—positive displacement, and
  - SCADA systems; and
- Maintain the following equipment:
  - Aeration basins,
  - Aeration systems (e.g., blowers, surface aerators, diffusers),
  - Air compressors,

- Analyzers (e.g., dissolved oxygen, pH, H_2S, ORP),
- Bar screens,
- Chlorination systems,
- Clarifiers/sedimentation basins,
- Dechlorination systems,
- Gates and valves,
- Generators,
- Hand tools,
- Heavy equipment,
- Instrumentation (e.g., flow, pressure, telemetry),
- Mixers,
- Motors,
- Ponds/lagoons,
- Power tools,
- Pumps—centrifugal,
- Pumps—positive displacement, and
- Suspended growth (e.g., activated sludge, MBR, SBR).

Class I Types of Knowledge Required to Perform Job Tasks

Types of Knowledge	Level of Knowledge
Chemical handling and storage	Basic
Chlorinators (e.g., gas, liquid)	Basic
Clarifiers	Basic
Comminuters	Basic
Conveyors	Basic
Dewatering equipment (e.g., centrifuges, presses, drying beds)	Basic
Documentation and recordkeeping	Basic
Electrical principles (e.g., troubleshooting breakers, relays, circuits)	Basic
Emergency preparedness	Basic
Flow measuring devices (e.g., Parshals flumes, mag meter, venturis)	Basic
Grit removal processes (e.g., grit chamber, aeration, cyclone)	Basic
Heavy equipment (e.g., operation, preventive maintenance)	Basic
Industry safety practices (e.g., PPE, confined spaces, fall arrest, lockout/tagout)	Basic
Maintenance practices (e.g., preventive, reactive, predictive)	Basic

Types of Knowledge	Level of Knowledge
Ozone generation equipment	Basic
Pneumatic principles (e.g., troubleshooting actuators, compressors, sprayers)	Basic
Primary treatment processes (e.g., ponds, sedimentation basins)	Basic
Principles of asset management (e.g., preventive, reactive, predictive maintenance)	Basic
Process control instrumentation (e.g., PLCs, SCADA, continuous online monitoring)	Basic
Screening technology (e.g., bar screens, microscreens)	Basic
Secondary treatment processes (e.g., activated sludge, MBR, SBR)	Basic
Solids treatment concepts (e.g., dewatering, digestion, thickening)	Basic
Tertiary treatment processes (e.g., media filtration, disinfection, post-aeration, reclaimed recharge)	Basic
Treatment equipment (e.g., pumps, motors, generators)	Basic

Class I Sample Questions

1. Which of the following is NOT a closed-pipe flowmeter?

    a) Orifice plate
    b) Propeller
    c) Magnetic
    d) Flume

2. Pressure measurement devices are used directly in water resource recovery facilities for maintaining proper

    a) pressures in lubrication and seal systems.
    b) flow measurement in a mechanical closed flowmeter.
    c) suction lift of a wet well pump.
    d) float operations.

3. If a motor starter trips a motor, leads can be disconnected and the windings checked, reading each phase to ground. If the volt-ohmmeter shows a low ohm reading, the phase is

    a) normal.
    b) grounded.
    c) looped.
    d) three-phase.

4.	A possible cause of low output capacity from a plunger pump could be

	a)	an air leak in suction piping.
	b)	the pump packing being too tight.
	c)	the pump stroke being too long.
	d)	insufficient or excessive lubrication.

5.	The MAIN reason that electrical distribution networks are grounded is to

	a)	prevent short-circuiting.
	b)	provide alternating current (AC) power.
	c)	provide direct current (DC) power.
	d)	prevent the development of unsafe conditions.

6.	Which component of anaerobic digester gas can form acid in a digester boiler?

	a)	Carbon dioxide
	b)	Hydrogen sulfide
	c)	Methane
	d)	Siloxanes

7.	The MOST important parameter to monitor on a chemical feed pump is

	a)	vibration.
	b)	flow rate.
	c)	day tank level.
	d)	pump run times.

8.	Which of the following is the recommended material for liquid chlorine piping?

	a)	Carbon steel
	b)	Copper
	c)	Chlorinated polyvinyl chloride (CPVC)
	d)	Rubber

9.	On a circular clarifier mechanism, which of the following pins is designed to fail in an over-torque situation before any structural damage occurs to the sludge collection mechanism?

	a)	Cotter
	b)	Fail safe
	c)	Torque
	d)	Shear

10. Misalignment between a motor and a pump could cause

   a) improved motor efficiency.
   b) shaft damage.
   c) excess capacity.
   d) excessive pump speed.

11. The most common cause of motor malfunction is

   a) bearing failure.
   b) dust and dirt.
   c) thermal overload.
   d) single phasing.

12. What is the primary reason to regularly exercise and lubricate a typically closed plug valve?

   a) To ensure that the operators know where it is located
   b) To prevent the material deposited in the "dead end" pipe from becoming septic
   c) To ensure that the valve operates correctly when it is needed
   d) To provide work for more operations staff

13. The primary maintenance concern of a float system location in a stilling well measuring raw wastewater is

   a) a change in temperature.
   b) high pH.
   c) accumulation of debris.
   d) low biochemical oxygen demand.

14. Before a piece of equipment or system is locked out and tagged,

   a) everyone must leave the area where the equipment is located.
   b) all personnel must be notified that the equipment or system will be locked out and tagged.
   c) laboratory personnel must be consulted to avoid loss of process efficiency.
   d) local emergency response agencies must be alerted.

15. If a grit cyclone has been observed to be vibrating during maintenance rounds, what task should be performed to correct this problem?

   a) Take the unit out of service and remove debris
   b) No action is required; vibration is normal for grit cyclones

c) Increase the grit concentration in the unit feed

d) Replace the cycle motor bearing

16. Which chemical can be used to detect chlorine leaks?

a) Ammonia

b) Hydrogen sulfide

c) Methane

d) Sulfur dioxide

17. If a magnetic flow meter with a span of 0 to 1000 gpm (0 to 63 L/s) has a specific accuracy of 2% of the span, the reading would be expected to be within what actual maximum flow?

a) 2 gpm (0.1 L/s)

b) 10 gpm (0.6 L/s)

c) 20 gpm (1.3 L/s)

d) 200 gpm (12.6 L/s)

18. Liquid sodium hypochlorite storage tanks should be protected from light and heat because

a) the heat and sunlight will increase the chlorine concentration of the chemical.

b) the heat and light will increase scaling.

c) the heat and light will have no effect of the chemical characteristics.

d) the heat and light will hasten the solution's deterioration.

19. When determining the stocking of spare parts for a primary settling tank and other facility equipment maintenance, personnel must take into account

a) color and size.

b) cost and manufacturer representative selection.

c) number of items and storage area size.

d) equipment fail occurrence and delivery time.

Answers

1.   **Answer: D**

A flume is an open-channel flow measurement device.

Reference: Water Environment Federation (2008) *Operation of Municipal Wastewater Treatment Plants*, 6th ed.; Manual of Practice No. 11; Water Environment Federation: Alexandria, Virginia; pp 7-11–7-13.

2. **Answer:** A

Maintaining proper pressures in lubrication and seal systems is a common direct use of pressure measurement devices in a water resource recovery facility.

Reference: Water Environment Federation (2008) *Operation of Municipal Wastewater Treatment Plants*, 6th ed.; Manual of Practice No. 11; Water Environment Federation: Alexandria, Virginia; p 7-15.

3. **Answer:** B

A low ohm reading may mean a phase is grounded.

Reference: Water Environment Federation (2008) *Operation of Municipal Wastewater Treatment Plants*, 6th ed.; Manual of Practice No. 11; Water Environment Federation: Alexandria, Virginia; p 10-34.

4. **Answer:** A

Plunger pumps do not use packing. If a pump's stroke is too long, it produces a higher output. Excessive lubrication would cause heat and friction, but would not directly affect output capacity. An air leak in the suction piping would cause at least partial loss of vacuum and reduce the amount of fluid pulled into the pump with each stroke.

Reference: Water Environment Federation (2008) *Operation of Municipal Wastewater Treatment Plants*, 6th ed.; Manual of Practice No. 11; Water Environment Federation: Alexandria, Virginia; p 8-63.

5. **Answer:** D

Safety is the primary purpose for grounding.

Reference: Water Environment Federation (2008) *Operation of Municipal Wastewater Treatment Plants*, 6th ed.; Manual of Practice No. 11; Water Environment Federation: Alexandria, Virginia; p 10-25.

6. **Answer:** B

Hydrogen sulfide is an extremely reactive compound that, when combined with water, forms an acid solution that is highly corrosive to pipelines.

Reference: Water Environment Federation (2008) *Operation of Municipal Wastewater Treatment Plants*, 6th ed.; Manual of Practice No. 11; Water Environment Federation: Alexandria, Virginia; p 30-38.

7. **Answer:** B

Chemical and liquid feeders, such as metering pumps, can have many operating problems that prevent the feeder from delivering the correct amount of chemical.

Reference: Water Environment Federation (2008) *Operation of Municipal Wastewater Treatment Plants*, 6th ed.; Manual of Practice No. 11; Water Environment Federation: Alexandria, Virginia; p 9-42.

8. **Answer:** C

Recommended materials of construction differ for chlorine gas and liquid sodium hypochlorite. Gaseous chlorine uses steel cylinders with carbon steel piping and valves. Liquid sodium hypochlorite systems use fiber-glass-reinforced plastic (FRP) or chlorinated polyvinyl chloride (CPVC) storage tanks, and piping and nonmetallic or plastic-lined valves.

Reference: Water Environment Federation (2008) *Operation of Municipal Wastewater Treatment Plants*, 6th ed.; Manual of Practice No. 11; Water Environment Federation: Alexandria, Virginia; p 9-22, Table 9.5.

9. **Answer:** D

A shear pin is a straight pin that will fail (break) when a certain load or stress is exceeded. The purpose of the pin is to protect equipment from damage caused by excessive loads or stress.

Reference: California State University, Sacramento (2008) *Operation of Wastewater Treatment Plants*, 7th ed.; California State University: Sacramento, California; Volume I, p 539.

10. **Answer:** B

Motor and pump misalignment can result in damaged bearings, broken shafts, and excessively worn or ruined gears.

Reference: California State University, Sacramento (2007) *Operation of Wastewater Treatment Plants*, 7th ed.; California State University: Sacramento, California; Volume II, p 429.

11. **Answer:** C

Thirty percent of motor malfunctions are caused by thermal overload.

Reference: California State University, Sacramento (2007) *Operation of Wastewater Treatment Plants*, 7th ed.; California State University: Sacramento, California; Volume II, p 390.

12. **Answer:** C

Valves need to be regularly lubricated and exercised to ensure ease of operation.

Reference: Water Environment Federation (2008) *Operation of Municipal Wastewater Treatment Plants*, 6th ed.; Manual of Practice No. 11; Water Environment Federation: Alexandria, Virginia; p 8-87.

13. Answer: C

Stilling wells are often used to create a calm area where the true water elevation can be measured without turbulence affecting the reading. Debris, solids, and ice need to be removed from these wells to prevent interference with accurate measurements.

Reference: Water Environment Federation (2008) *Operation of Wastewater Treatment Plants,* 6th ed.; Manual of Practice No. 11; Water Environment Federation: Alexandria, Virginia; p 7-18.

14. Answer: B

All personnel must be notified that the equipment is being locked out and tagged.

Reference: California State University, Sacramento (2007) *Operation of Wastewater Treatment Plants,* 7th ed.; California State University: Sacramento, California; Volume II, p 274.

15. Answer: A

If debris blocks the cyclone, the smooth centrifugal flow pattern is disrupted and turbulence occurs in the unit. This turbulence causes vibration. The condition can be easily corrected by taking the unit out of service and removing the debris.

Reference: Water Environment Federation (2008) *Operation of Municipal Wastewater Treatment Plants,* 6th ed.; Manual of Practice No. 11; Water Environment Federation: Alexandria, Virginia; Table 18-17.

16. Answer: A

Gaseous ammonia will detect any chlorine leak by creating a white ammonium chloride cloud when puffed near the leak. Never pour ammonia solution or water onto a chlorine leak because hypochlorous and hydrochloric acid will form, which will corrode the steel pipe or fitting.

Reference: Water Environment Federation (2008) *Operation of Municipal Wastewater Treatment Plants,* 6th ed.; Manual of Practice No. 11; Water Environment Federation: Alexandria, Virginia; p 26-32.

17. Answer: C

$$1000 \text{ gpm}\left[\frac{2}{100}\right] = 20 \text{ gpm}$$

$$63 \text{ L/s}\left[\frac{2}{100}\right] = 1.26 \text{ L/s}$$

Reference: Water Environment Federation (2008) *Operation of Municipal Wastewater Treatment Plants,* 6th ed.; Manual of Practice No. 11; Water Environment Federation: Alexandria, Virginia; p 7-8.

18. Answer: D

Sodium hypochlorite storage tanks should be protected against light and heat because these conditions will hasten the solution's deterioration, thus reducing the strength (available chlorine strength) of the solution.

Reference: Water Environment Federation (2008) *Operation of Municipal Wastewater Treatment Plants*, 6th ed.; Manual of Practice No. 11; Water Environment Federation: Alexandria, Virginia; pp 26-36–26-37.

19. Answer: D

When determining the stocking of spare parts, personnel must take into account the fail occurrence of a piece of equipment and the delivery time of a piece of equipment.

Reference: Water Environment Federation (2008) *Operation of Municipal Wastewater Treatment Plants*, 6th ed.; Manual of Practice No. 11; Water Environment Federation: Alexandria, Virginia; p 19-36.

Class II Job Tasks

- Calibrate meters (e.g., flow, pressure sensors);
- Follow safety rules and guidelines when working with chemical equipment;
- Follow safety rules and guidelines when working with mechanical equipment;
- Monitor flowmeters;
- Monitor telemetry systems;
- Perform basic electrical troubleshooting;
- Perform preventive maintenance on equipment;
- Inspect the following equipment:
  - Aeration basins,
  - Aeration systems (e.g., blowers, surface aerators, diffusers),
  - Aerobic digesters,
  - Air compressors,
  - Analyzers (e.g., dissolved oxygen, pH, H_2S, ORP),
  - Attached growth/fixed film (e.g., RBC, trickling filter),
  - Bar screens,
  - Chemical feed systems (e.g., polymer, ferric),
  - Chlorination systems,
  - Clarifiers/sedimentation basins,
  - Dechlorination systems,
  - Disinfection equipment (e.g., UV, ozone),
  - Flow equalization systems,

- Gates and valves,
- Generators,
- Grit removal processes,
- Hand tools,
- Heavy equipment,
- Hoists and cranes,
- Instrumentation (e.g., flow, pressure, telemetry),
- Mechanical dewatering equipment (e.g., presses, centrifuges),
- Mixers,
- Motors,
- Power tools,
- Pumps—centrifugal,
- Pumps—positive displacement,
- SCADA systems,
- Solids thickening processes (e.g., DAF, belt, rotary drum), and
- Suspended growth (e.g., activated sludge, MBR, SBR); and
- Maintain the following equipment:
  - Aeration basins,
  - Aeration systems (e.g., blowers, surface aerators, diffusers),
  - Aerobic digesters,
  - Air compressors,
  - Analyzers (e.g., dissolved oxygen, pH, H_2S, ORP),
  - Attached growth/fixed film (e.g., RBC, trickling filter),
  - Bar screens,
  - Chemical feed systems (e.g., polymer, ferric),
  - Chlorination systems,
  - Clarifiers/sedimentation basins,
  - Dechlorination systems,
  - Disinfection equipment (e.g., UV, ozone),
  - Gates and valves,
  - Generators,
  - Grit removal processes,
  - Hand tools,
  - Heavy equipment,
  - Hoists and cranes,
  - Instrumentation (e.g., flow, pressure, telemetry),
  - Mechanical dewatering equipment (e.g., presses, centrifuges),
  - Mixers,
  - Motors,

  ○ Power tools,
  ○ Pumps—centrifugal,
  ○ Pumps—positive displacement,
  ○ SCADA systems, and
  ○ Suspended growth (e.g., activated sludge, MBR, SBR).

Class II Types of Knowledge Required to Perform Job Tasks

Types of Knowledge	Level of Knowledge
Aeration principles (e.g., mixing, mechanical, diffusers)	Basic
Chemical handling and storage	Basic
Chlorinators (e.g., gas, liquid)	Basic
Clarifiers	Basic
Comminuters	Basic
Conveyors	Basic
Dewatering equipment (e.g., centrifuges, presses, drying beds)	Basic
Documentation and recordkeeping	Basic
Electrical principles (e.g., troubleshooting breakers, relays, circuits)	Basic
Emergency preparedness	Basic
Flow measuring devices (e.g., Parshall flumes, mag meter, venturis)	Basic
Grit removal processes (e.g., grit chamber, aeration, cyclone)	Basic
Heavy equipment (e.g., operation, preventive maintenance)	Basic
Industry safety practices (e.g., PPE, confined spaces, fall arrest, lockout/tagout)	Basic
Maintenance practices (e.g., preventive, reactive, predictive)	Basic
Ozone generation equipment	Basic
Pneumatic principles (e.g., troubleshooting actuators, compressors, sprayers)	Basic
Primary treatment processes (e.g., ponds, sedimentation basins)	Basic
Principles of asset management (e.g., preventive, reactive, predictive maintenance)	Basic
Process control instrumentation (e.g., PLCs, SCADA, continuous online monitoring)	Basic
Screening technology (e.g., bar screens, microscreens)	Basic
Secondary treatment processes (e.g., activated sludge, MBR, SBR)	Basic
Solids treatment concepts (e.g., dewatering, digestion, thickening)	Basic
Tertiary treatment processes (e.g., media filtration, disinfection, post-aeration, reclaimed recharge)	Basic
Treatment equipment (e.g., pumps, motors, generators)	Basic

Class II Sample Questions

1. A _____ meter is used to measure flow in a pipe by creating a measurable pressure drop across the device.

    a) propeller
    b) differential head
    c) differential suction
    d) magnetic

2. Which type of speed measurement consists of a stator and rotor?

    a) Tachometer generator
    b) Noncontact frequency generator
    c) Magnetic proximity sensor
    d) Capacity proximity generator

3. Which preliminary treatment device has the largest steel bar spacing in the channel?

    a) In-channel grinder/comminutor
    b) Trash rack
    c) Bar screen
    d) Fine screen

4. Typical bar screen spacing is

    a) 0.25 to 0.75 in. (6.3 to 19 mm)
    b) 0.75 to 2.0 in. (19 to 51 mm)
    c) 2 to 6 in. (51 to 152 mm)
    d) Greater than 6 in. (152 mm)

5. There are different methods used to start the cleaning cycles for mechanical screens. Which method is NOT used to control the cleaning cycle of a mechanical screen?

    a) Programmable logic controllers (PLCs)
    b) Timing devices
    c) Differential level
    d) Pressure sensor

6. What area of a boiler system is most likely to need maintenance?

    a) Combustion chamber
    b) Flue

    c) Heat exchanger
    d) Water supply

7. You are going to add ferric chloride solution at the headworks of the plant. What type of feeder would **NOT** be used?

    a) Centrifugal
    b) Diaphragm
    c) Gravimetric
    d) Peristaltic

8. During rounds, the operator notices that there is a broken chain in the rectangular primary tank. Which of the following would **NOT** be a cause of the broken scraper chain?

    a) Excessive loading on the sludge collector
    b) Debris caught on the primary tank weirs
    c) Ice build-up in the tank
    d) Improper shear pin sizing and flight alignment

9. On a gravity belt thickener, chicanes are used for what purpose?

    a) To distribute spray water onto the belt
    b) To keep the belt properly tensioned
    c) To assist with the removal of thickened sludge from the end of the belt
    d) To clear paths on the belt to assist with filtrate drainage

10. At a minimum, when the temperature is below freezing, how often should you check the pressure and vacuum relief valves on an anaerobic digester?

    a) Daily
    b) Weekly
    c) Monthly
    d) Semiannually

11. How often should critical motors undergo bearing inspection, testing of windings, and cleaning?

    a) Monthly
    b) Annually
    c) Every 5 years
    d) Every 10 years

12. When buildup of deposits is observed in a rotameter, what should you do?

   a) Clean the tube and float
   b) Increase flow to flush
   c) Increase pressure to the flush
   d) Nothing as these are normal conditions

13. How often should the secondary clarifier drive unit be checked for condensation?

   a) Weekly
   b) Monthly
   c) Annually
   d) Never

14. For what is "CMMS" an acronym?

   a) Computerized mechanical microscopic system
   b) Computational mechanical maintenance system
   c) Computerized maintenance management system
   d) Carbonaceous management mechanical support

15. A sodium hypochlorite system had been operating properly for a while, but suddenly the sodium hypochlorite feed pump stops pumping. What could be the reason?

   a) The pump was recently rebuilt and tested by the manufacturer.
   b) There is too much chemical entering the pump.
   c) The pump is gas locking so the pump needs to be switched to a suction lift pump.
   d) The pump is gas locking so the pump suction conditions must be checked to ensure that the pump has a flooded suction.

16. After performing maintenance on a positive-displacement pump, which of the following tasks should be performed before starting the pump?

   a) Ensure the valve(s) on the suction and discharge side of the pump are opened.
   b) Ensure the seal water is turned off to the pump.
   c) Ensure the pump motor is reading the full load amperage.
   d) Ensure the pump motor is disconnected.

17. In the equation $E = IR$, which variable represents current?

   a) E
   b) I

   c)   *R*

   d)   None represent current

18. Bucket- and chain-type mechanically cleaned grit removal units need regular checking and cleaning of the _____ at the bottom of the buckets to ensure that water drains properly from the buckets.

   a)   flap valves

   b)   drain holes

   c)   shear pins

   d)   wear shoes

Answers

1. **Answer:** B

    Differential head meter is used to measure flow in a closed pipe.

    Reference: Water Environment Federation (2008) *Operation of Municipal Wastewater Treatment Plants*, 6th ed.; Manual of Practice No. 11; Water Environment Federation: Alexandria, Virginia; p 7-11.

2. **Answer:** A

    A tachometer is the only device with both of these parts.

    Reference: Water Environment Federation (2008) *Operation of Municipal Wastewater Treatment Plants*, 6th ed.; Manual of Practice No. 11; Water Environment Federation: Alexandria, Virginia; p 7-27.

3. **Answer:** B

    Bar racks are the coarsest screens used.

    Reference: Water Environment Federation (2008) *Operation of Municipal Wastewater Treatment Plants*, 6th ed.; Manual of Practice No. 11; Water Environment Federation: Alexandria, Virginia; p 18-3.

4. **Answer:** B

    Other sizes besides ¼ and ¾ in. are defined as different specialized screens.

    Reference: Water Environment Federation (2008) *Operation of Municipal Wastewater Treatment Plants*, 6th ed.; Manual of Practice No. 11; Water Environment Federation: Alexandria, Virginia; p 18-4.

5. **Answer:** D

Mechanically cleaned screens use PLCs, timing devices, and level controls to start a cleaning cycle. Mechanically cleaned screens will use one or a combination of methods to start a cleaning cycle.

Reference: Water Environment Federation (2008) *Operation of Municipal Wastewater Treatment Plants*, 6th ed.; Manual of Practice No. 11; Water Environment Federation: Alexandria, Virginia; p 18-7.

6. **Answer:** A

Boilers should be inspected as recommended by the boiler manufacturer, paying particular attention to the **<u>COMBUSTION AREA</u>**, which is susceptible to corrosion.

Reference: Water Environment Federation (2008) *Operation of Municipal Wastewater Treatment Plants*, 6th ed.; Manual of Practice No. 11; Water Environment Federation: Alexandria, Virginia; p 30-36.

7. **Answer:** C

Reference: Water Environment Federation (2008) *Operation of Municipal Wastewater Treatment Plants*, 6th ed.; Manual of Practice No. 11; Water Environment Federation: Alexandria, Virginia; p 9-28.

8. **Answer:** B

Debris caught on the weirs will not cause the sludge collector to break.

Reference: Water Environment Federation (2008) *Operation of Municipal Wastewater Treatment Plants*, 6th ed.; Manual of Practice No. 11; Water Environment Federation: Alexandria, Virginia; Table 19-4, p 19-33.

9. **Answer:** D

Chicanes (plows) are used to clear paths on the belt to help drain the filtrate and they are used to turn the sludge rows to improve drainage.

Reference: Water Environment Federation (2008) *Operation of Municipal Wastewater Treatment Plants*, 6th ed.; Manual of Practice No. 11; Water Environment Federation: Alexandria, Virginia; p 29-42.

10. **Answer:** A

These two valves should be checked at least every 6 months for proper operation, and daily under freezing conditions.

Reference: California State University, Sacramento (2007) *Operation of Wastewater Treatment Plants*, 7th ed.; California State University: Sacramento, California; Volume II, p 162.

11. **Answer:** C

Critical motors should be tested for integrity of insulation. Additionally, every 5 to 6 years, the motors should be sent out for testing of the windings; bearing inspection; and cleaning, revarnishing, and baking.

Reference: Water Environment Federation (2008) *Operation of Municipal Wastewater Treatment Plants*, 6th ed.; Manual of Practice No. 11; Water Environment Federation: Alexandria, Virginia; p 10-28.

12. **Answer:** A

Clean the rotometer when deposits are noticed inside the glass rotometer tube or if the float sticks. Be sure that the rotometer and associated equipment are completely dry before reassembling. If moisture is present, chlorine will combine with it to form hypochlorous and hydrochloric acid, which will cause corrosion to occur.

Reference: Water Environment Federation (2008) *Operation of Municipal Wastewater Treatment Plants*, 6th ed.; Manual of Practice No. 11; Water Environment Federation: Alexandria, Virginia; p 26-34.

13. **Answer:** A

The secondary clarifier drive should be checked weekly for condensation.

Reference: Water Environment Federation (2008) *Operation of Municipal Wastewater Treatment Plants*, 6th ed.; Manual of Practice No. 11; Water Environment Federation: Alexandria, Virginia; p 20-216.

14. **Answer:** C

Reference: Water Environment Federation (2008) *Operation of Municipal Wastewater Treatment Plants*, 6th ed.; Manual of Practice No. 11; Water Environment Federation: Alexandria, Virginia; p 12-23.

15. **Answer:** D

Sodium hypochlorite can release gas bubbles that accumulate in the suction piping to the feed pumps, which may cause air binding of the feed pumps. Maintain a flooded suction or air-release mechanism to ensure proper operation.

Reference: Water Environment Federation (2008) *Operation of Municipal Wastewater Treatment Plants*, 6th ed.; Manual of Practice No. 11; Water Environment Federation: Alexandria, Virginia; p 26-36.

16. **Answer:** A

Positive-displacement pumps should not be operated against a closed discharge valve. Before operating a solids pump for the first time, operators should ensure that valves are open on both the suction and discharge sides of the pump. In addition to general pump maintenance activities, operators should seriously consider high-pressure protection because positive-displacement pumps will always deliver flow if the pump is running. Running against a closed or partially closed valve can result in unacceptably high pressures.

Reference: Water Environment Federation (2008) *Operation of Municipal Wastewater Treatment Plants*, 6th ed.; Manual of Practice No. 11; Water Environment Federation: Alexandria, Virginia; p 8-58.

17. **Answer: B**

E = voltage, I = current, and R = resistance.

Reference: Water Environment Federation (2008) *Operation of Municipal Wastewater Treatment Plants*, 6th ed.; Manual of Practice No. 11; Water Environment Federation: Alexandria, Virginia; p 10-6.

18. **Answer: B**

Periodic inspection and cleaning of the drain holes is required on the chain and bucket grit collection systems to ensure that water is flowing out of the bucket so that a drier grit is placed into the dumpster.

Reference: Water Environment Federation (2008) *Operation of Municipal Wastewater Treatment Plants*, 6th ed.; Manual of Practice No. 11; Water Environment Federation: Alexandria, Virginia; p 18-16.

Class III Job Tasks

- Calibrate meters (e.g., flow, pressure sensors);
- Follow safety rules and guidelines when working with chemical equipment;
- Follow safety rules and guidelines when working with mechanical equipment;
- Monitor flowmeters;
- Monitor telemetry systems;
- Perform basic electrical troubleshooting;
- Perform preventive maintenance on equipment;
- Inspect the following equipment:
  - Aeration basins,
  - Aeration systems (e.g., blowers, surface aerators, diffusers),
  - Air compressors,
  - Anaerobic digesters,
  - Analyzers (e.g., dissolved oxygen, pH, H_2S, ORP),
  - Bar screens,

- ○ Chemical feed systems (e.g., polymer, ferric),
- ○ Chlorination systems,
- ○ Clarifiers/sedimentation basins,
- ○ Conveyors,
- ○ Dechlorination systems,
- ○ Disinfection equipment (e.g., UV, ozone),
- ○ Flow equalization systems,
- ○ Gates and valves,
- ○ Generators,
- ○ Grit removal processes,
- ○ Hand tools,
- ○ Heavy equipment,
- ○ Hoists and cranes,
- ○ Instrumentation (e.g., flow, pressure, telemetry),
- ○ Mechanical dewatering equipment (e.g., presses, centrifuges),
- ○ Mixers,
- ○ Motors,
- ○ Odor control devices (e.g., biofilters, scrubbers),
- ○ Power tools,
- ○ Pumps—centrifugal,
- ○ Pumps—positive displacement,
- ○ SCADA systems,
- ○ Solids thickening processes (e.g., DAF, belt, rotary drum), and
- ○ Suspended growth (e.g., activated sludge, MBR, SBR); and
- • Maintain the following equipment:
  - ○ Aeration basins,
  - ○ Aeration systems (e.g., blowers, surface aerators, diffusers),
  - ○ Air compressors,
  - ○ Anaerobic digesters,
  - ○ Analyzers (e.g., dissolved oxygen, pH, H_2S, ORP),
  - ○ Bar screens,
  - ○ Chemical feed systems (e.g., polymer, ferric),
  - ○ Chlorination systems,
  - ○ Clarifiers/sedimentation basins,
  - ○ Conveyors,
  - ○ Dechlorination systems,
  - ○ Disinfection equipment (e.g., UV, ozone),
  - ○ Flow equalization systems,
  - ○ Gates and valves,

- ○ Generators,
- ○ Grit removal processes,
- ○ Hand tools,
- ○ Heavy equipment,
- ○ Hoists and cranes,
- ○ Instrumentation (e.g., flow, pressure, telemetry),
- ○ Mechanical dewatering equipment (e.g., presses, centrifuges),
- ○ Mixers,
- ○ Motors,
- ○ Power tools,
- ○ Pumps—centrifugal,
- ○ Pumps—positive displacement,
- ○ SCADA systems,
- ○ Solids thickening processes (e.g., DAF, belt, rotary drum), and
- ○ Suspended growth (e.g., activated sludge, MBR, SBR).

Class III Types of Knowledge Required to Perform Job Tasks

Types of Knowledge	Level of Knowledge
Aeration principles (e.g., mixing, mechanical, diffusers)	Intermediate
Chemical handling and storage	Intermediate
Chlorinators (e.g., gas, liquid)	Intermediate
Clarifiers	Intermediate
Comminuters	Intermediate
Conveyors	Intermediate
Dewatering equipment (e.g., centrifuges, presses, drying beds)	Intermediate
Documentation and recordkeeping	Intermediate
Electrical principles (e.g., troubleshooting breakers, relays, circuits)	Intermediate
Emergency preparedness	Intermediate
Flow measuring devices (e.g., Parshall flumes, mag meter, venturis)	Intermediate
Grit removal processes (e.g., grit chamber, aeration, cyclone)	Intermediate
Heavy equipment (e.g., operation, preventive maintenance)	Intermediate
Hydraulic principles (e.g., mass flow balance, detention time, loading, velocity, HRT)	Basic
Industry safety practices (e.g., PPE, confined spaces, fall arrest, lockout/tagout)	Intermediate
Maintenance practices (e.g., preventive, reactive, predictive)	Intermediate

Types of Knowledge	Level of Knowledge
Ozone generation equipment	Intermediate
Pneumatic principles (e.g., troubleshooting actuators, compressors, sprayers)	Intermediate
Primary treatment processes (e.g., ponds, sedimentation basins)	Intermediate
Principles of asset management (e.g., preventive, reactive, predictive maintenance)	Intermediate
Process control instrumentation (e.g., PLCs, SCADA, continuous online monitoring)	Intermediate
Screening technology (e.g., bar screens, microscreens)	Intermediate
Secondary treatment processes (e.g., activated sludge, MBR, SBR)	Intermediate
Solids treatment concepts (e.g., dewatering, digestion, thickening)	Intermediate
Tertiary treatment processes (e.g., media filtration, disinfection, post-aeration, reclaimed recharge)	Intermediate
Treatment equipment (e.g., pumps, motors, generators)	Intermediate

Class III Sample Questions

1. If the rake on a mechanically cleaned screen is inoperable and there is no visible problem, the issue is likely a(n) _____.

    a) blockage
    b) broken chain
    c) electrical issue
    d) full dumpster

2. Resistance temperature detectors (RTDs) are most commonly used on _____ ambient-range temperatures.

    a) higher
    b) lower
    c) Neither a nor b
    d) Both a and b

3. A level sensor is being used to measure a wet well liquid level and the reading is 4 mA. What is the current level of a wet well that is 5 ft wide by 5 ft long by 10 ft deep (1.5 m wide by 1.5 m long by 3 m deep)? Assume that 1 ft is the lowest liquid level.

    a) 1 ft (0.3 m)
    b) 5 ft (1.7 m)
    c) 8 ft (2.7 m)
    d) 10 ft (3.3 m)

4. A mesophilic anaerobic digestion system uses a heating hot water system that contains a two heating loop system, referred to as the primary and secondary loops. The secondary hot water loop is designed to typically send hot water to the heat exchangers at a temperature of _____ °F.

   a) 95 °F (35 °C)
   b) 150 °F (66 °C)
   c) 200 °F (93 °C)
   d) 250 °F (121 °C)

5. When using a gravimetric vibrating trough feeder, one of the routine maintenance requirements is to

   a) check for caking.
   b) check for foaming.
   c) grease the wobble motor weekly.
   d) wash the trough daily.

6. What can cause a blockage in a chlorinator atmospheric vent line?

   a) Calcium deposits
   b) Field mice
   c) Insects
   d) Magnesium deposits

7. To keep an even flow pattern through a circular clarifier, what should be checked and adjusted on a yearly basis?

   a) Blades/scrapers
   b) Skimmers
   c) Spray bars
   d) Weirs

8. Which area of the comminutor should be checked if there are "ropes" of rags surrounding the drum of a comminutor?

   a) Cutter head
   b) Drum screen
   c) Inlet gate
   d) Outlet gate

9. Dosing polymer at a high rate onto a belt filter press can cause a condition known as
    _____ to show up in the belt filter press sludge on the gravity zone.

    a) clumping
    b) belt blinding
    c) roller slippage
    d) sludge deflocculating

10. Given the following:

    -   Aerobic digester dissolved oxygen level is 3.5 mg/L,
    -   All blowers are running,
    -   pH is 7.2,
    -   Temperature is 19 °C,
    -   Specific oxygen uptake rate (SOUR) is 1.2 mg/L O_2/g/h,
    -   High surface turbulence, and
    -   The tank is exhibiting high rates of foaming,

    What is first step you should take to reduce the foaming?

    a) Add an acid
    b) Add a caustic
    c) Reduce aeration
    d) Reduce wasting

11. When using a draft tube mixer on an anaerobic digester, which bearing is most likely
    to fail?

    a) Thrust
    b) Top
    c) Shaft
    d) Radial

12. A 1-in.-diameter shaft with a runout greater than _____ indicates a bent
    shaft.

    a) 0.000 2 in. (0.005 mm)
    b) 0.002 in. (0.05 mm)
    c) 0.02 in. (0.5 mm)
    d) 0.2 in. (5 mm)

13. When cleaning an electric motor with compressed air, what is the maximum recommended pressure you should use?

    a) 10 psi (69 kPa)
    b) 25 psi (172 kPa)
    c) 50 psi (345 kPa)
    d) 100 psi (690 kPa)

14. What is meant by a compound loop control for a chlorination system?

    a) The chlorine feed system is controlled by the facility influent flowmeter.
    b) The chlorine feed system is controlled by a facility flowmeter and chlorine residual analyzer.
    c) The chlorine system is controlled by the facility flowmeter, chlorine tank pH probe, and temperature.
    d) The chlorine feed system is controlled by the chlorine residual analyzer and the level in the chlorine contact tank.

15. If a pneumatic bubbler system for flow measurement becomes plugged, you should **NOT** use the following cleaning method:

    a) Chemical
    b) Overpressure
    c) Wire rodding
    d) Vacuum suction

16. Before doing maintenance on the lamps of a UV disinfection system, the power to the lamps must be turned off so exposed lamps out of the water do not cause what?

    a) Burns to the eyes and skin
    b) Hair loss and discolored fingernails
    c) Stress because of the lamps being pulled out
    d) Bleeding and scars

17. In a rectangular primary sedimentation tank, the collector mechanism has a noisy chain drive. What could be the problem?

    a) The shear pin is too small
    b) There is an oil leak in the gearbox
    c) A worn chain and/or sprocket
    d) The sludge pump is operating too frequently

18. Motors require regular inspection and maintenance. Insulation failures could occur when the motor

    a) is operated below its normal operating temperature.
    b) is operated below its full-load amperage reading.
    c) is subjected to deposits of dust and other foreign material.
    d) has a service factor of 1.2.

19. Maintenance is required on a pump at a water resource recovery facility. The parts for the repair cost $6500. Two maintenance personnel work on the pump for 6 hours each. An electrician worked on the pump problem for 3 hours. The labor rates for the municipality for these employees are $49/hour for the mechanics and $55/hour for the electrician. How much did it cost for the municipality to repair this pump?

    a) $6500
    b) $6959
    c) $7253
    d) $7418

Answers

1. **Answer: C**

    A blockage, a broken chain, and a full dumpster are all visible issues; an electrical issue is not.

    Reference: Water Environment Federation (2008) *Operation of Municipal Wastewater Treatment Plants*, 6th ed.; Manual of Practice No. 11; Water Environment Federation: Alexandria, Virginia; p 18-23.

2. **Answer: B**

    Resistance temperature detectors are used on lower ambient temperature ranges.

    Reference: Water Environment Federation (2008) *Operation of Municipal Wastewater Treatment Plants*, 6th ed.; Manual of Practice No. 11; Water Environment Federation: Alexandria, Virginia; p 7-23.

3. **Answer: A**

    Electronic instruments are measured where a 4-mA signal would typically indicate a low liquid level, low pump speed, or closed valve.

    Reference: Water Environment Federation (2008) *Operation of Municipal Wastewater Treatment Plants*, 6th ed.; Manual of Practice No. 11; Water Environment Federation: Alexandria, Virginia; p 7-38.

4.   **Answer:** B

One loop is used for the solids heat exchangers, with a recommended temperature of 66 °C (150 °F) to minimize fouling of the heat exchanger surfaces. The second loop supports the boiler, which should be maintained at a temperature of a least 82 °C (180 °F) to avoid the formation of sulfuric acid.

Reference: Water Environment Federation (2008) *Operation of Municipal Wastewater Treatment Plants*, 6th ed.; Manual of Practice No. 11; Water Environment Federation: Alexandria, Virginia; p 30-36.

5.   **Answer:** A

Care must be taken so the chemical does not cake in the hopper and stop feeding into the trough. In addition, caking on the trough will prevent an even flow of chemical, which could change the output volume.

Reference: California State University, Sacramento (2006) *Advanced Waste Treatment*, 5th ed.; California State University: Sacramento, California; p 370.

6.   **Answer:** C

Place a screen over the end of the pipe to keep insects from building a nest in the pipe and clogging it.

Reference: California State University, Sacramento (2007) *Operation of Wastewater Treatment Plants*, 7th ed.; California State University: Sacramento, California; Volume II, p 336.

7.   **Answer:** D

Weirs should be kept free of debris and level to help prevent short-circuiting.

Reference: California State University, Sacramento (2008) *Operation of Wastewater Treatment Plants*, 7th ed.; California State University: Sacramento, California; Volume I, p 131.

8.   **Answer:** A

Daily, or several times a day, check for "ropes" of rags hanging from the slotted drums or U-shaped bars. The presence of this type of debris indicates that the cutters may be worn or out of adjustment.

Reference: California State University, Sacramento (2008) *Operation of Wastewater Treatment Plants*, 7th ed.; California State University: Sacramento, California; Volume I, p 80.

9.   **Answer:** B

When too much polymer is added to the belt filter press, small white dots will appear in the sludge on the gravity zone. If this occurs, reduce the polymer dosage rate.

Another common problem that is encountered with belt filter presses is if the polymer dosage rate is too low. If this occurs, sludge will spill off the press under the side skirts.

Reference: California State University, Sacramento (2006) *Advanced Waste Treatment*, 5th ed.; California State University: Sacramento, California; p 244.

10. **Answer: C**

Excessive turbulence in the presence of either a non-biodegradable detergent or some filamentous bacteria can cause severe foaming. Because the dissolved oxygen concentration is higher than necessary and the digester is functioning well, the first thing to do would be to reduce the aeration rate, which will reduce the turbulence and see if that improves the foaming issue.

Reference: California State University, Sacramento (2006) *Advanced Waste Treatment*, 5th ed.; California State University: Sacramento, California; p 216.

11. **Answer: C**

The draft tube units are subject to shaft bearing failure caused by the abrasiveness of sludge and corrosion by hydrogen sulfide in the digester gas.

Reference: California State University, Sacramento (2008) *Operation of Wastewater Treatment Plants*, 7th ed.; California State University: Sacramento, California; Volume I, p 179.

12. **Answer: B**

Runout should be checked. A reading higher than 0.002 in. (0.05 mm) indicates a bent shaft or a shaft that is eccentrically machined.

Reference: California State University, Sacramento (2007) *Operation of Wastewater Treatment Plants*, 7th ed.; California State University: Sacramento, California; Volume II, p 400.

13. **Answer: C**

Blow dirt out with clean, dry, compressed air at 30 to 50 psi (207 to 245 kPa).

Reference: California State University, Sacramento (2007) *Operation of Wastewater Treatment Plants*, 7th ed.; California State University: Sacramento, California; Volume II, p 366.

14. **Answer: B**

The compound loop system controls the chlorine feed rate by using a facility flow signal to set the dosage and then uses a signal from the chlorine residual analyzer to readjust the dosage rate to the proper requirement.

Reference: Water Environment Federation (2008) *Operation of Municipal Wastewater Treatment Plants*, 6th ed.; Manual of Practice No. 11; Water Environment Federation: Alexandria, Virginia; p 7-42.

15. **Answer:** B

Do not attempt to pressurize the system at higher than normal operating pressure for cleaning. Such action will damage internal parts.

Reference: California State University, Sacramento (2008) *Operation of Wastewater Treatment Plants,* 7th ed.; California State University: Sacramento, California; Volume I, p 456.

16. **Answer:** A

Reference: California State University, Sacramento (2008) *Operation of Wastewater Treatment Plants,* 7th ed.; California State University: Sacramento, California; Volume I, p 417.

17. **Answer:** C

A noisy chain drive can be from a loose chain, worn sprocket, rubbing parts, or misalignment.

Reference: Water Environment Federation (2008) *Operation of Municipal Wastewater Treatment Plants,* 6th ed.; Manual of Practice No. 11; Water Environment Federation: Alexandria, Virginia; Table 19-4, p 19-33.

18. **Answer:** C

Reference: California State University, Sacramento (2007) *Operation of Wastewater Treatment Plants,* 7th ed.; California State University: Sacramento, California; Volume II, p 391.

19. **Answer:** C

The cost of the pump repair was
    Cost for parts: $6500.00
    Mechanic labor rate: $49/hour
    Electrician labor rate: $55/hour
    Number of mechanics working on pump: 2
    Number of electricians working on pump: 1
    Mechanic work hours on pump: 6 hours per mechanic
    Electrician work hours on pump: 3 hours per electrician

Total project cost, dollars = Parts cost + Mechanic labor + Electrician labor
$$= \$6500 + (2 \text{ mechanics})(6 \text{ hours})(\$49/\text{hour}) +$$
$$(1 \text{ electrician})(3 \text{ hours})(\$55/\text{hour})$$
$$= \$6500 + \$588 + \$165$$
$$= \$7253$$

Reference: Water Environment Federation (2008) *Operation of Municipal Wastewater Treatment Plants,* 6th ed.; Manual of Practice No. 11; Water Environment Federation: Alexandria, Virginia; p 3-32.

Class IV Job Tasks

- Calibrate meters (e.g., flow, pressure sensors);
- Follow safety rules and guidelines when working with chemical equipment;
- Follow safety rules and guidelines when working with mechanical equipment;
- Monitor flowmeters;
- Monitor telemetry systems;
- Inspect the following equipment:
  - Aeration basins,
  - Aeration systems (e.g., blowers, surface aerators, diffusers),
  - Air compressors,
  - Anaerobic digesters,
  - Analyzers (e.g., dissolved oxygen, pH, H_2S, ORP),
  - Chemical feed systems (e.g., polymer, ferric),
  - Clarifiers/sedimentation basins,
  - Disinfection equipment (e.g., UV, ozone),
  - Filtration and exchange units (e.g., sand, membranes),
  - Generators,
  - Grit removal processes,
  - Instrumentation (e.g., flow, pressure, telemetry),
  - Mechanical dewatering equipment (e.g., presses, centrifuges),
  - Odor control devices (e.g., biofilters, scrubbers),
  - Pumps—centrifugal,
  - Pumps—positive displacement,
  - SCADA systems,
  - Solids thickening processes (e.g., DAF, belt, rotary drum), and
  - Suspended growth (e.g., activated sludge, MBR, SBR); and
- Maintain the following equipment:
  - Aeration basins,
  - Aeration systems (e.g., blowers, surface aerators, diffusers),
  - Air compressors,
  - Anaerobic digesters,
  - Analyzers (e.g., dissolved oxygen, pH, H_2S, ORP),
  - Bar screens,
  - Chemical feed systems (e.g., polymer, ferric),
  - Clarifiers/sedimentation basins,
  - Disinfection equipment (e.g., UV, ozone),
  - Generators,
  - Grit removal processes,
  - Instrumentation (e.g., flow, pressure, telemetry),
  - Mechanical dewatering equipment (e.g., presses, centrifuges),

    ○   Odor control devices (e.g., biofilters, scrubbers),

    ○   Pumps—centrifugal,

    ○   Pumps—positive displacement,

    ○   SCADA systems, and

    ○   Suspended growth (e.g., activated sludge, MBR, SBR).

Class IV Types of Knowledge Required to Perform Job Tasks

Types of Knowledge	Level of Knowledge
Aeration principles (e.g., mixing, mechanical, diffusers)	Intermediate
Chemical handling and storage	Advanced
Chlorinators (e.g., gas, liquid)	Advanced
Clarifiers	Advanced
Comminuters	Advanced
Conveyors	Advanced
Dewatering equipment (e.g., centrifuges, presses, drying beds)	Advanced
Documentation and recordkeeping	Advanced
Electrical principles (e.g., troubleshooting breakers, relays, circuits)	Intermediate
Emergency preparedness	Advanced
Flow measuring devices (e.g., Parshall flumes, mag meter, venturis)	Intermediate
Grit removal processes (e.g., grit chamber, aeration, cyclone)	Advanced
Heavy equipment (e.g., operation, preventive maintenance)	Intermediate
Hydraulic principles (e.g., mass flow balance, detention time, loading, velocity, HRT)	Intermediate
Industry safety practices (e.g., PPE, confined spaces, fall arrest, lockout/tagout)	Advanced
Maintenance practices (e.g., preventive, reactive, predictive)	Advanced
Ozone generation equipment	Intermediate
Pneumatic principles (e.g., troubleshooting actuators, compressors, sprayers)	Intermediate
Primary treatment processes (e.g., ponds, sedimentation basins)	Advanced
Principles of asset management (e.g., preventive, reactive, predictive maintenance)	Advanced
Process control instrumentation (e.g., PLCs, SCADA, continuous online monitoring)	Intermediate
Screening technology (e.g., bar screens, microscreens)	Intermediate
Secondary treatment processes (e.g., activated sludge, MBR, SBR)	Advanced
Solids treatment concepts (e.g., dewatering, digestion, thickening)	Advanced
Tertiary treatment processes (e.g., media filtration, disinfection, post-aeration, reclaimed recharge)	Advanced
Treatment equipment (e.g., pumps, motors, generators)	Advanced

Class IV Sample Questions

1. Which of the following is (are) not a component of a magnetic flowmeter?

    a) Electrode
    b) Pipe section
    c) Magnetic coils
    d) Straightening vanes

2. When a mechanical aerator is operating below the manufacturer's recommendation and a wave pattern is established that causes the impeller to alternately be over- and under-submerged, this condition is called

    a) overloading.
    b) hydraulic surging.
    c) impeller fouling.
    d) proper operation of a mechanical aerator.

3. The condition when a column of water that typically travels downward in an aerator draft tube reverses and starts flowing upward is called

    a) surging.
    b) hydraulic overloading.
    c) flooding.
    d) impeller fouling.

4. Most chemical metering pumps should have what differential pressure across the valves?

    a) <5 psi (<34 kPa)
    b) 5–10 psi (34–69 kPa)
    c) 10–20 psi (69–138 kPa)
    d) >20 psi (>138 kPa)

5. When cleaning a sodium hypochlorite pump and effluent water strainer, what type of gloves should be worn?

    a) No gloves are necessary
    b) Leather
    c) Kevlar
    d) Rubber

6. Given the following information about a primary clarifier:

    -   Sludge is very thin,
    -   Sludge collector motor running,
    -   The flow over the weirs appears normal, and
    -   Sludge collector is moving

    What is the most probable problem?

    a) Broken collector chain
    b) Broken shear pin
    c) Excessive sludge pumping
    d) Inadequate sludge pumping

7. The screening belt conveyor belt is not tracking properly. The most likely cause is

    a) the drive pulley has stretched.
    b) the drainage ports are blocked with debris.
    c) the tail pulley is misaligned.
    d) the idler sprocket is damaged.

8. A thermographic infrared inspection of the electrical distribution system should be performed

    a) while the system is deenergized.
    b) after nitrogen gas has been applied to avoid overheating.
    c) while the system is energized and under full load.
    d) once the fuse bus has been uploaded.

9. Which of the following common precipitates is the **LEAST** desirable in the maintenance of an anaerobic digestion system?

    a) Bicarbonate
    b) Carbonate
    c) Struvite
    d) Vivianite

10. Overloading an electric motor will decrease the life of its insulation. A continuous overload of 12% will decrease the insulation life approximately

    a) 25%.
    b) 50%.

    c) 75%.

    d) 95%.

11. For a magnetic flowmeter to read correctly, it should be located _____ pipe diameters downstream from any valves.

    a) 2

    b) 5

    c) 10

    d) 15

12. Spiral heat exchangers used for anaerobic digestion require regular inspection to prevent plugging caused by the narrow concentric spiral passages within the unit. Inlet and outlet pressure gauge readings should be taken daily. A plug in the heat exchanger is evident when

    a) both pressure gauges are reading close together with little pressure drop.

    b) both pressure gauges are reading zero when the pump is operating.

    c) the inlet pressure gauge has a zero reading and the discharge pressure gauge has a positive reading.

    d) there is a significant increase in pressure drop across the heat exchanger.

13. On a secondary clarifier drive unit, it is recommended that the torque switches be checked

    a) daily.

    b) weekly.

    c) monthly.

    d) yearly.

Answers

1.   **Answer: D**

   Only straightening vanes are not part of a magnetic flowmeter.

   Reference: Water Environment Federation (2008) *Operation of Municipal Wastewater Treatment Plants*, 6th ed.; Manual of Practice No. 11; Water Environment Federation: Alexandria, Virginia; p 7-13.

2.   **Answer: B**

   "Hydraulic surging" is what is being defined.

Reference: Water Environment Federation (2008) *Operation of Municipal Wastewater Treatment Plants*, 6th ed.; Manual of Practice No. 11; Water Environment Federation: Alexandria, Virginia; p 20-110.

3. **Answer:** C

Flooding is the reverse of flow in a draft tube aerator.

Reference: Water Environment Federation (2008) *Operation of Municipal Wastewater Treatment Plants*, 6th ed.; Manual of Practice No. 11; Water Environment Federation: Alexandria, Virginia; p 20-111.

4. **Answer:** B

Most metering pumps should have a 5- to 10-psig (34- to 69-kPa) differential across the valves.

Reference: Water Environment Federation (2008) *Operation of Municipal Wastewater Treatment Plants*, 6th ed.; Manual of Practice No. 11; Water Environment Federation: Alexandria, Virginia; p 9-40.

5. **Answer:** D

Water strainers on chlorinators frequently clog and require attention. Water strainers may be cleaned by being flushed with water, or, if they are badly fouled, they may be cleaned with dilute hydrochloric acid followed by a water rinse. Occasionally, pump a 5% solution of hydrochloric acid to prevent carbonate scale from accumulating in the pump. Flush the pump with water before and after using acid. Rubber gloves should be worn to prevent chemical burns during maintenance activities. In addition, protective eyewear should be used.

Reference: Water Environment Federation (2008) *Operation of Municipal Wastewater Treatment Plants*, 6th ed.; Manual of Practice No. 11; Water Environment Federation: Alexandria, Virginia; p 26-36.

6. **Answer:** C

The blanket thickness and solids concentration in a primary clarifier are inversely proportional to the underflow pumping rate. Higher pumping rates will result in lower blankets and lower solids concentrations. Lower pumping rates will result in deeper blankets and higher solids concentrations.

Reference: Water Environment Federation (2008) *Operation of Municipal Wastewater Treatment Plants*, 6th ed.; Manual of Practice No. 11; Water Environment Federation: Alexandria, Virginia; p 19-30, Table 19.3.

7. **Answer:** C

    Tail pulley is misaligned.

    Reference: Water Environment Federation (2008) *Operation of Municipal Wastewater Treatment Plants*, 6th ed.; Manual of Practice No. 11; Water Environment Federation: Alexandria, Virginia; p 18-24, Table 18.2.

8. **Answer:** C

    Thermographic infrared inspection should be conducted while the system is energized and under full load.

    Reference: Water Environment Federation (2008) *Operation of Municipal Wastewater Treatment Plants*, 6th ed.; Manual of Practice No. 11; Water Environment Federation: Alexandria, Virginia; p 10-33.

9. **Answer:** C

    The digestion process can produce crystalline precipitates. Common precipitates include struvite ($MgNH_4PO_4$), vivianite ($Fe_3(PO_4)_2$), and calcium carbonate ($CaCO_3$). Vivianite is the desired product when iron is added to control struvite. Adding iron sequesters the phosphate, which removes one of the ingredients necessary for struvite formation. Less common precipitates include various calcium complexes such as calcium carbonate.

    Reference: Water Environment Federation (2008) *Operation of Municipal Wastewater Treatment Plants*, 6th ed.; Manual of Practice No. 11; Water Environment Federation: Alexandria, Virginia; p 30-61.

10. **Answer:** A

    A continuous 12% overload cuts insulation life to one-quarter of its design life.

    Reference: California State University, Sacramento (2007) *Operation of Wastewater Treatment Plants*, 7th ed.; California State University: Sacramento, California; Volume II, p 392.

11. **Answer:** B

    Electromagnetic meters require full-pipe flow at all times and a clear length of 5 pipe diameters upstream and downstream. Doppler meters require a pipe material that will allow penetration of an ultrasonic signal, horizontal installation, and a clear length of 10 pipe diameters upstream and 5 diameters downstream. Venturi meters require full-pipe flow at all times and a clear length of pipe 20 throat diameters upstream and 10 throat diameters downstream.

    Reference: Water Environment Federation (2008) *Operation of Municipal Wastewater Treatment Plants*, 6th ed.; Manual of Practice No. 11; Water Environment Federation: Alexandria, Virginia; p 25-51.

12. **Answer:** D

Reference: Water Environment Federation (2008) *Operation of Municipal Wastewater Treatment Plants*, 6th ed.; Manual of Practice No. 11; Water Environment Federation: Alexandria, Virginia; p 30-74, Table 30.12.

13. **Answer:** B

Reference: Water Environment Federation (2008) *Operation of Municipal Wastewater Treatment Plants*, 6th ed.; Manual of Practice No. 11; Water Environment Federation: Alexandria, Virginia; p 20-216.

Equipment Operation

Class I Job Tasks

- Analyze data to evaluate and adjust equipment;
- Check filters for proper operation;
- Conduct wastewater pipe repairs;
- Follow safety rules and guidelines when working with chemical equipment;
- Follow safety rules and guidelines when working with mechanical equipment;
- Follow standard operating procedures (SOPs);
- Monitor lift stations to ensure equipment is operating properly;
- Monitor motor control center; and
- Operate the following:
  - Aeration basins,
  - Aeration systems (e.g., blowers, surface aerators, diffusers),
  - Aerobic digesters,
  - Air compressors,
  - Analyzers (e.g., dissolved oxygen, pH, H_2S, ORP),
  - Bar screens,
  - Chemical feed systems (e.g., polymer, ferric),
  - Chlorination systems,
  - Clarifiers/sedimentation basins,
  - Conveyors,
  - Dechlorination systems,
  - Gates and valves,
  - Generators,
  - Hand tools,
  - Heavy equipment,
  - Hoists and cranes,
  - Instrumentation (e.g., flow, pressure, telemetry),
  - Mixers,
  - Motors,
  - Ponds/lagoons,
  - Power tools,
  - Pumps—centrifugal,
  - Pumps—positive displacement, and
  - SCADA systems.

Class I Types of Knowledge Required to Perform Job Tasks

Types of Knowledge	Level of Knowledge
Aeration principles (e.g., mixing, mechanical, diffusers)	Basic
Chemical handling and storage	Basic
Chlorinators (e.g., gas, liquid)	Basic
Clarifiers	Basic
Comminuters	Basic
Conveyors	Basic
Dewatering equipment (e.g., centrifuges, presses, drying beds)	Basic
Electrical principles (e.g., troubleshooting breakers, relays, circuits)	Basic
Emergency preparedness	Basic
Flow measuring devices (e.g., Parshall flumes, mag meter, venturis)	Basic
Grit removal processes (e.g., grit chamber, aeration, cyclone)	Basic
Heavy equipment (e.g., operation, preventive maintenance)	Basic
Hydraulic principles (e.g., mass flow balance, detention time, loading, velocity, HRT)	Basic
Industry safety practices (e.g., PPE, confined spaces, fall arrest, lockout/tagout)	Basic
Maintenance practices (e.g., preventive, reactive, predictive)	Basic
Ozone generation equipment	Basic
Pneumatic principles (e.g., troubleshooting actuators, compressors, sprayers)	Basic
Primary treatment processes (e.g., ponds, sedimentation basins)	Basic
Principles of asset management (e.g., preventive, reactive, predictive maintenance)	Basic
Process control instrumentation (e.g., PLCs, SCADA, continuous online monitoring)	Basic
Screening technology (e.g., bar screens, microscreens)	Basic
Secondary treatment processes (e.g., activated sludge, MBR, SBR)	Basic
Solids treatment concepts (e.g., dewatering, digestion, thickening)	Basic
Tertiary treatment processes (e.g., media filtration, disinfection, post-aeration, reclaimed recharge)	Basic
Treatment equipment (e.g., pumps, motors, generators)	Basic

Class I Sample Questions

1. A sudden reduction of pump flow can be caused by

    a) debris.
    b) a volute.
    c) a leaking seal.
    d) a discharge valve stuck open.

2. A pump impeller is used to create a

   a) seal.
   b) direct displacement force.
   c) centrifugal force that moves a liquid.
   d) strong connection between a pump and its metallic base.

3. An air gap backflow preventer is used to

   a) prime a pump.
   b) separate the pump parts from the pumped fluid.
   c) keep wastewater from entering the potable water system.
   d) provide oxygen to an aerating basin.

4. Electrical transformers are used to

   a) change electricity into amps.
   b) increase or decrease voltage.
   c) transfer electricity into wastewater.
   d) protect plant workers from electrocution.

5. When using a positive-displacement pump, what should you never do?

   a) Pump fluid uphill
   b) Lubricate the pump while it is running
   c) Supply the mechanical seal with water
   d) Run with the discharge valve closed

6. In wastewater, _____ pumps are most commonly used to pump influent, effluent, return activated sludge, waste activated sludge, and so on.

   a) peripheral
   b) propeller
   c) screw lift
   d) centrifugal

Answers

1. **Answer:** A

   Pump troubleshooting can be difficult; however, most of the time debris is the culprit. Manufacturer operation and maintenance manuals can also be quite helpful in troubleshooting.

Reference: Water Environment Federation (2008) *Operation of Municipal Wastewater Treatment Plants*, 6th ed.; Manual of Practice No. 11; Water Environment Federation: Alexandria, Virginia; pp 8-15–8-17 and 8-44.

2. **Answer:** C

A pump impeller creates a pressure drop when it spins. This change in pressure causes water to be taken in and expelled from the pump.

Reference: Water Environment Federation (2008) *Operation of Municipal Wastewater Treatment Plants*, 6th ed.; Manual of Practice No. 11; Water Environment Federation, Alexandria: Virginia; pp 8-1 and 8-15–8-17.

3. **Answer:** C

Air gaps between equipment and pipes prevent wastewater bacteria from entering the potable water supply. Regular maintenance and inspection are required.

Reference: Water Environment Federation (2008) *Operation of Municipal Wastewater Treatment Plants*, 6th ed.; Manual of Practice No. 11; Water Environment Federation: Alexandria, Virginia; pp 11-3–11-4.

4. **Answer:** B

Transformers take power from the power company and change it so it can be used to run machinery. Once in the plant, other transformers are used to change the power to operate lights and smaller motors and equipment. The following voltages are found in most treatment plants: 220 V, 480 V, and higher voltages.

Reference: Water Environment Federation (2008) *Operation of Municipal Wastewater Treatment Plants*, 6th ed.; Manual of Practice No. 11; Water Environment Federation: Alexandria, Virginia; p 10-44.

5. **Answer:** D

Running a positive-displacement pump against a closed valve will severely damage the pump and piping and represents a significant safety hazard.

Reference: Water Environment Federation (2008) *Operation of Municipal Wastewater Treatment Plants*, 6th ed.; Manual of Practice No. 11; Water Environment Federation: Alexandria, Virginia; pp 8-15–8-17, 8-48, and 8-57.

6. **Answer:** D

All of the pumps may appear in the wastewater process; however, the centrifugal pump is the most common.

Reference: Water Environment Federation (2008) *Operation of Municipal Wastewater Treatment Plants*, 6th ed.; Manual of Practice No. 11; Water Environment Federation: Alexandria, Virginia; p 8-29.

Class II Job Tasks

- Analyze data to evaluate and adjust equipment;
- Check filters for proper operation;
- Conduct wastewater pipe repairs;
- Follow safety rules and guidelines when working with chemical equipment;
- Follow safety rules and guidelines when working with mechanical equipment;
- Follow standard operating procedures (SOPs);
- Monitor lift stations to ensure equipment is operating properly;
- Monitor motor control center;
- Transport biosolids offsite for disposal/reuse; and
- Operate the following:
  - Aeration basins,
  - Aeration systems (e.g., blowers, surface aerators, diffusers),
  - Aerobic digesters,
  - Air compressors,
  - Analyzers (e.g., dissolved oxygen, pH, H_2S, ORP),
  - Attached growth/fixed film (e.g., RBC, trickling filter),
  - Bar screens,
  - Chemical feed systems (e.g., polymer, ferric),
  - Chlorination systems,
  - Clarifiers/sedimentation basins,
  - Dechlorination systems,
  - Disinfection equipment (e.g., UV, ozone),
  - Filtration and exchange units (e.g., sand, membranes),
  - Flow equalization systems,
  - Gates and valves,
  - Generators,
  - Grit removal processes,
  - Hand tools,
  - Heavy equipment,
  - Hoists and cranes,
  - Instrumentation (e.g., flow, pressure, telemetry),
  - Mechanical dewatering equipment (e.g., presses, centrifuges),
  - Mixers,
  - Motors,
  - Odor control devices (e.g., biofilters, scrubbers),

- ○ Power tools,
- ○ Pumps—centrifugal,
- ○ Pumps—positive displacement,
- ○ SCADA systems,
- ○ Solids thickening processes (e.g., DAF, belt, rotary drum), and
- ○ Suspended growth (e.g., activated sludge, MBR, SBR).

Class II Types of Knowledge Required to Perform Job Tasks

Types of Knowledge	Level of Knowledge
Aeration principles (e.g., mixing, mechanical, diffusers)	Basic
Chemical handling and storage	Basic
Chlorinators (e.g., gas, liquid)	Basic
Clarifiers	Basic
Comminuters	Basic
Conveyors	Basic
Dewatering equipment (e.g., centrifuges, presses, drying beds)	Basic
Electrical principles (e.g., troubleshooting breakers, relays, circuits)	Basic
Emergency preparedness	Basic
Flow measuring devices (e.g., Parshall flumes, mag meter, venturis)	Basic
Grit removal processes (e.g., grit chamber, aeration, cyclone)	Basic
Heavy equipment (e.g., operation, preventive maintenance)	Basic
Hydraulic principles (e.g., mass flow balance, detention time, loading, velocity, HRT)	Basic
Industry safety practices (e.g., PPE, confined spaces, fall arrest, lockout/tagout)	Basic
Maintenance practices (e.g., preventive, reactive, predictive)	Basic
Ozone generation equipment	Basic
Pneumatic principles (e.g., troubleshooting actuators, compressors, sprayers)	Basic
Primary treatment processes (e.g., ponds, sedimentation basins)	Basic
Principles of asset management (e.g., preventive, reactive, predictive maintenance)	Basic
Process control instrumentation (e.g., PLCs, SCADA, continuous online monitoring)	Basic
Screening technology (e.g., bar screens, microscreens)	Basic
Secondary treatment processes (e.g., activated sludge, MBR, SBR)	Basic
Solids treatment concepts (e.g., dewatering, digestion, thickening)	Basic
Tertiary treatment processes (e.g., media filtration, disinfection, post-aeration, reclaimed recharge)	Basic
Treatment equipment (e.g., pumps, motors, generators)	Basic

Class II Sample Questions

1.   If an aeration tank requires more air, a simple solution is to

     a)   increase the speed of the blower.
     b)   make larger holes in the piping.
     c)   slightly close the blower suction valve.
     d)   put more water in the tank so the air has more water to mix with.

2.   A way to determine the amount of dissolved oxygen in a tank is to

     a)   use a handheld pH meter.
     b)   look at the size of the bubbles in the tank.
     c)   measure the air gap between the air bubbles and water.
     d)   use a handheld dissolved oxygen meter.

3.   A common way to measure flow in a pipe is to

     a)   use a magnetic flowmeter.
     b)   measure the speed of ping pong balls in the pipe.
     c)   measure how far the water sprays when it leaves the pipe.
     d)   measure the water level over a V-notch weir.

4.   Diffuser-based air systems consist of

     a)   a few small air tanks spread over a large area.
     b)   a blower and a pipe distribution system that are used to bubble air into water.
     c)   different air concentrations in a tank.
     d)   high-speed rotating mixers.

5.   What equipment or actions are typically used to control air flow from centrifugal blowers?

     a)   Throttle the inlet valves or use variable-frequency drives
     b)   Place the blower into surge condition
     c)   Trim the blower impeller
     d)   Throttle the discharge valve and variable-speed drives

Answers

1.   **Answer:** A

     Increasing the speed of a blower is a practical way to introduce more air to a tank, but it does have its limits. Air piping can be blown apart if you exceed the design pressure of the piping.

Reference: Water Environment Federation (2008) *Operation of Municipal Wastewater Treatment Plants*, 6th ed.; Manual of Practice No. 11; Water Environment Federation: Alexandria, Virginia; pp 20-3–20-34.

2. **Answer:** D

Portable handheld dissolved oxygen meters are handy in the field. They are easy to operate and can be carried to a tank to perform measurements.

Reference: Water Environment Federation (2008) *Operation of Municipal Wastewater Treatment Plants*, 6th ed.; Manual of Practice No. 11; Water Environment Federation: Alexandria, Virginia; pp 20-38–20-43.

3. **Answer:** A

A magnetic flowmeter is the most common form of measurement for flows in a pipe.

Reference: Water Environment Federation (2008) *Operation of Municipal Wastewater Treatment Plants*, 6th ed.; Manual of Practice No. 11; Water Environment Federation: Alexandria, Virginia; p 7-11.

4. **Answer:** B

The system is made up of a blower, controls, air piping, headers on the tank, and some type of opening in the bottom of the tank. The openings can be coarse, medium, or fine pores. Coarse diffusers can be very large (up to 4 in. by 4 in. openings) and fine-pore diffusers can have openings the size of pinholes.

Reference: Water Environment Federation (2008) *Operation of Municipal Wastewater Treatment Plants*, 6th ed.; Manual of Practice No. 11; Water Environment Federation: Alexandria, Virginia; pp 20-32 and 20-84.

5. **Answer:** A

Variable-speed drives or inlet valves are the best way to control the output of centrifugal blowers. Any type of control on the discharge side will cause significant damage to equipment.

Reference: Water Environment Federation (2008) *Operation of Municipal Wastewater Treatment Plants*, 6th ed.; Manual of Practice No. 11; Water Environment Federation: Alexandria, Virginia; p 20-82.

Class III Job Tasks

- Analyze data to evaluate and adjust equipment;
- Check filters for proper operation;

- Conduct wastewater pipe repairs;
- Follow safety rules and guidelines when working with chemical equipment;
- Follow safety rules and guidelines when working with mechanical equipment;
- Follow standard operating procedures (SOPs);
- Monitor lift stations to ensure equipment is operating properly;
- Monitor motor control center;
- Transport biosolids offsite for disposal/reuse; and
- Operate the following:
  - Aeration basins,
  - Aeration systems (e.g., blowers, surface aerators, diffusers),
  - Air compressors,
  - Anaerobic digesters,
  - Analyzers (e.g., dissolved oxygen, pH, H_2S, ORP),
  - Bar screens,
  - Chemical feed systems (e.g., polymer, ferric),
  - Chlorination systems,
  - Clarifiers/sedimentation basins,
  - Conveyors,
  - Dechlorination systems,
  - Disinfection equipment (e.g., UV, ozone),
  - Flow equalization systems,
  - Gates and valves,
  - Generators,
  - Grit removal processes,
  - Hand tools,
  - Heavy equipment,
  - Hoists and cranes,
  - Instrumentation (e.g., flow, pressure, telemetry),
  - Mechanical dewatering equipment (e.g., presses, centrifuges),
  - Mixers,
  - Motors,
  - Odor control devices (e.g., biofilters, scrubbers),
  - Power tools,
  - Pumps—centrifugal,
  - Pumps—positive displacement,
  - SCADA systems,
  - Solids thickening processes (e.g., DAF, belt, rotary drum), and
  - Suspended growth (e.g., activated sludge, MBR, SBR).

Class III Types of Knowledge Required to Perform Job Tasks

Types of Knowledge	Level of Knowledge
Aeration principles (e.g., mixing, mechanical, diffusers)	Intermediate
Chemical handling and storage	Intermediate
Chlorinators (e.g., gas, liquid)	Intermediate
Clarifiers	Intermediate
Comminuters	Intermediate
Conveyors	Intermediate
Dewatering equipment (e.g., centrifuges, presses, drying beds)	Intermediate
Electrical principles (e.g., troubleshooting breakers, relays, circuits)	Intermediate
Emergency preparedness	Intermediate
Flow measuring devices (e.g., Parshall flumes, mag meter, venturis)	Intermediate
Grit removal processes (e.g., grit chamber, aeration, cyclone)	Intermediate
Heavy equipment (e.g., operation, preventive maintenance)	Intermediate
Hydraulic principles (e.g., mass flow balance, detention time, loading, velocity, HRT)	Intermediate
Industry safety practices (e.g., PPE, confined spaces, fall arrest, lockout/tagout)	Intermediate
Maintenance practices (e.g., preventive, reactive, predictive)	Intermediate
Ozone generation equipment	Intermediate
Pneumatic principles (e.g., troubleshooting actuators, compressors, sprayers)	Intermediate
Primary treatment processes (e.g., ponds, sedimentation basins)	Intermediate
Principles of asset management (e.g., preventive, reactive, predictive maintenance)	Intermediate
Process control instrumentation (e.g., PLCs, SCADA, continuous online monitoring)	Intermediate
Screening technology (e.g., bar screens, microscreens)	Intermediate
Secondary treatment processes (e.g., activated sludge, MBR, SBR)	Intermediate
Solids treatment concepts (e.g., dewatering, digestion, thickening)	Intermediate
Tertiary treatment processes (e.g., media filtration, disinfection, post-aeration, reclaimed recharge)	Intermediate
Treatment equipment (e.g., pumps, motors, generators)	Intermediate

Class III Sample Questions

1.   To help ensure reliable operation, emergency generators should be exercised

   a)   twice a week without load.

   b)   daily with some plant equipment running (under load).

    c)   monthly with normal plant equipment running.

    d)   only when needed.

2.   To safely reduce the flowrate of a centrifugal pump,

    a)   decrease the motor speed.

    b)   decrease the total dynamic head.

    c)   close the discharge valve.

    d)   increase the rotational speed of the impeller.

3.   Pipes used for chemicals that are acidic or basic should be

    a)   compatible with the chemical being pumped.

    b)   stainless steel.

    c)   polyvinyl chloride.

    d)   aluminum.

4.   What is typically used as a safety device for machinery that may experience unusually high torque while running?

    a)   Lockout/tagout system

    b)   High-pressure sensors

    c)   Air compressors

    d)   Shear pins

5.   The formation and collapse of gas pockets or bubbles on the blade of a centrifugal pump impeller can cause

    a)   air binding of the impeller.

    b)   compression of water on the leading edges of the impeller.

    c)   hydraulic overload of the impeller.

    d)   physical damage on the impeller.

6.   A water resource recovery facility has three trickling filters with rock media. Two trickling filters are in-service and one is out of service as a standby. One of the in-service trickling filters is experiencing ponding on the filter surface. Which of the following is **NOT** a solution to correct the performance?

    a)   Flood the filter for 24 hours.

    b)   Increase the hydraulic loading on the filter.

    c)   Shut down the filter until the media dries out.

    d)   Increase the organic loading on the filter.

7. When a new pure oxygen plant starts up, who is the **BEST** source for initial safety training and equipment operation and maintenance of the pure oxygen equipment?

    a) Facility manager
    b) Lead operator
    c) Equipment/oxygen provider
    d) Facility safety officer

Answers

1. **Answer:** C

    Emergency generators run more efficiently when they are exercised often. Running generators once per month lubricates the generator properly, reduces moisture buildup and corrosion, and reduces emissions. If they are not exercised regularly, they may have difficulty starting. Testing should allow the generator to reach its operating temperature.

    Reference: Water Environment Federation (2008) *Operation of Municipal Wastewater Treatment Plants*, 6th ed.; Manual of Practice No. 11; Water Environment Federation: Alexandria, Virginia; p 10-28.

2. **Answer:** A

    Reducing the motor speed of a pump will safely decrease the pump output as normal mode of operation if the pump is connected to a variable-frequency drive unit. If a constant-speed pump output needs to be consistently reduced, then the diameter of the impeller can be reduced. Typically, an impeller diameter change is used to improve pump efficiency and is considered a permanent change. Reducing total dynamic head will increase the discharge rate. Closing a discharge valve is extremely inefficient and causes high pressures inside the pump. Increasing the rotational speed will increase discharge rates (affinity laws).

    Reference: Water Environment Federation (2008) *Operation of Municipal Wastewater Treatment Plants*, 6th ed.; Manual of Practice No. 11; Water Environment Federation: Alexandria, Virginia; pp 8-11 and 8-24.

3. **Answer:** A

    Always be careful to check process chemicals and their compatibility with piping materials. High concentrations of acids, bases, or gaseous chlorine can cause pipes to fail if they are not compatible with each other and may also cause injuries.

    Reference: Water Environment Federation (2008) *Operation of Municipal Wastewater Treatment Plants*, 6th ed.; Manual of Practice No. 11; Water Environment Federation: Alexandria, Virginia; p 24-12.

4.   **Answer: D**

A shear pin is a device that will break under high-torque conditions and separate the motor from the drive unit, effectively shutting the system down before serious damage occurs. Shear pins are effective, simple, and very common, especially with low-speed, high-torque devices. The concept is simple: two couplings, face to face, share a common pin. If high torque develops, the pin breaks and the couplings separate before any of the equipment is damaged. It is important to replace the shear pin with the exact same pin recommended by the manufacturer.

Reference: Water Environment Federation (2008) *Operation of Municipal Wastewater Treatment Plants*, 6th ed.; Manual of Practice No. 11; Water Environment Federation: Alexandria, Virginia; pp 19-24–19-36.

5.   **Answer: D**

Cavitation is a phenomenon common in centrifugal pumps that occurs when the pump attempts to discharge more flow than it is pulling in through suction. Reduced internal pressure causes gases to expand and form bubbles around the impeller; when the bubbles implode, it potentially can cause severe damage to the pump.

Reference: Water Environment Federation (2008) *Operation of Municipal Wastewater Treatment Plants*, 6th ed.; Manual of Practice No. 11; Water Environment Federation: Alexandria, Virginia; pp 8-13–8-15.

6.   **Answer: D**

Ponding on the surface of a trickling filter is often due to organic overloading. Rapid growth of the biofilm can plug the openings in the media. Reducing the organic loading to the trickling filter will reduce the growth in the media bed and open up some void spaces for the water to pass through the media. Do not increase the organic loading to the trickling filter. Flooding the filter, increasing the hydraulic loading rate to the trickling filter, and shutting down the trickling filter are some solutions to ponding problems.

Reference: Water Environment Federation (2008) *Operation of Municipal Wastewater Treatment Plants*, 6th ed.; Manual of Practice No. 11; Water Environment Federation: Alexandria, Virginia; p 21-24.

7.   **Answer: C**

The equipment/oxygen provider is the best source for safety training and equipment operation and maintenance of the pure oxygen equipment.

Reference: Water Environment Federation (2013) *Safety, Health, and Security in Wastewater Systems*, 6th ed.; Manual of Practice No. 1; Water Environment Federation: Alexandria, Virginia; p 94.

Class IV Job Tasks

- Analyze data to evaluate and adjust equipment;
- Check filters for proper operation;
- Follow safety rules and guidelines when working with chemical equipment;
- Follow safety rules and guidelines when working with mechanical equipment;
- Follow standard operating procedures (SOPs);
- Monitor lift stations to ensure equipment is operating properly;
- Monitor motor control center; and
- Operate the following:
  - Aeration basins,
  - Aeration systems (e.g., blowers, surface aerators, diffusers),
  - Air compressors,
  - Anaerobic digesters,
  - Analyzers (e.g., DO, pH, H_2S, ORP),
  - Chemical feed systems (e.g., polymer, ferric),
  - Clarifiers/sedimentation basins,
  - Disinfection equipment (e.g., UV, ozone),
  - Filtration and exchange units (e.g., sand, membranes),
  - Generators,
  - Grit removal processes,
  - Instrumentation (e.g., flow, pressure, telemetry),
  - Mechanical dewatering equipment (e.g., presses, centrifuges),
  - Motors,
  - Odor control devices (e.g., biofilters, scrubbers),
  - Pumps—centrifugal,
  - Pumps—positive displacement,
  - SCADA systems,
  - Solids thickening processes (e.g., DAF, belt, rotary drum), and
  - Suspended growth (e.g., activated sludge, MBR, SBR).

Class IV Types of Knowledge Required to Perform Job Tasks

Types of Knowledge	Level of Knowledge
Aeration principles (e.g., mixing, mechanical, diffusers)	Advanced
Chemical handling and storage	Advanced
Chlorinators (e.g., gas, liquid)	Advanced
Clarifiers	Advanced

Types of Knowledge	Level of Knowledge
Comminuters	Advanced
Conveyors	Advanced
Dewatering equipment (e.g., centrifuges, presses, drying beds)	Advanced
Electrical principles (e.g., troubleshooting breakers, relays, circuits)	Advanced
Emergency preparedness	Advanced
Flow measuring devices (e.g., Parshall flumes, mag meter, venturis)	Advanced
Grit removal processes (e.g., grit chamber, aeration, cyclone)	Advanced
Heavy equipment (e.g., operation, preventive maintenance)	Advanced
Hydraulic principles (e.g., mass flow balance, detention time, loading, velocity, HRT)	Advanced
Industry safety practices (e.g., PPE, confined spaces, fall arrest, lockout/tagout)	Advanced
Maintenance practices (e.g., preventive, reactive, predictive)	Advanced
Ozone generation equipment	Advanced
Pneumatic principles (e.g., troubleshooting actuators, compressors, sprayers)	Advanced
Primary treatment processes (e.g., ponds, sedimentation basins)	Advanced
Principles of asset management (e.g., preventive, reactive, predictive maintenance)	Advanced
Process control instrumentation (e.g., PLCs, SCADA, continuous online monitoring)	Advanced
Screening technology (e.g., bar screens, microscreens)	Advanced
Secondary treatment processes (e.g., activated sludge, MBR, SBR)	Advanced
Solids treatment concepts (e.g., dewatering, digestion, thickening)	Advanced
Tertiary treatment processes (e.g., media filtration, disinfection, post-aeration, reclaimed recharge)	Advanced
Treatment equipment (e.g., pumps, motors, generators)	Advanced

Class IV Sample Questions

1. A basic pump curve plots the relationship between the system head and

    a) fluid temperature.
    b) fluid flow.
    c) impeller type.
    d) casing type.

2. Work has been completed on a motor for a centrifugal pump. The pump is now running, but generating little pressure and flow. What is the most likely cause?

a) The speed of the impeller is too high.

b) The pump is overly primed.

c) There is a water leak in the seal.

d) The impeller is rotating the wrong way.

3. What is considered an acceptable solids capture rate from solids thickening processes such as belt presses and centrifuges?

a) 65%

b) 75%

c) 85%

d) 95%

4. If a variable-speed drive with a maximum speed of 60 Hz is running at 35 Hz, how fast is the motor (maximum speed of 1200 rpm) spinning in revolutions per minute?

a) 7

b) 70

c) 700

d) 42 000

5. A centrifuge is fed 70 dry tons of sludge, the centrifuge discharges 65 dry tons of sludge, and the centrate generates 2.4 mil. gal (9 ML) of 500-ppm liquid that is returned back to the facility. Calculate the solids capture percent (%) of the centrifuge.

a) 83%

b) 85%

c) 93%

d) 98%

6. Calculate the annual operating cost of a pump given a flow of 950 gpm (60 L/s); total head of 45 ft (14 m); pump efficiency of 72%; motor efficiency of 94%; pump run time of 3 hours/day, 7 days/week; and energy cost of $0.055/kWh.

a) $102.41

b) $714.71

c) $1464.25

d) $7168.76

7. Which type of flow meter requires low suspended solids and debris to measure accurately?

a) Magnetic

b) Ultrasonic

    c) Mechanical

    d) Differential head

Answers

1. **Answer: B**

    Head curves can be difficult to interpret; however, the basic head curve plots total system head vs fluid flow.

    Reference: Water Environment Federation (2008) *Operation of Municipal Wastewater Treatment Plants*, 6th ed.; Manual of Practice No. 11; Water Environment Federation: Alexandria, Virginia; p 8-8.

2. **Answer: D**

    If three-phase power is wired incorrectly, the pump will run backwards at a reduced rate.

    Reference: Water Environment Federation (2008) *Operation of Municipal Wastewater Treatment Plants*, 6th ed.; Manual of Practice No. 11; Water Environment Federation: Alexandria, Virginia; p 10-3.

3. **Answer: D**

    The goal is to capture solids. Solids capture from centrifuges, filters, and dissolved air flotation should be 95% or greater. If solids capture is less than 90%, highly loaded side streams are returned to the plant, causing problems in other areas.

    Reference: Water Environment Federation (2008) *Operation of Municipal Wastewater Treatment Plants*, 6th ed.; Manual of Practice No. 11; Water Environment Federation: Alexandria, Virginia; pp 29-16–29-20.

4. **Answer: C**

    Solve as a simple ratio.

    $$\frac{35 \text{ Hz}}{60 \text{ Hz}} = \frac{\text{unknown rpm}}{1200 \text{ rpm}}$$

    700 rpm

    Reference: Water Environment Federation (2008) *Operation of Municipal Wastewater Treatment Plants*, 6th ed.; Manual of Practice No. 11; Water Environment Federation: Alexandria, Virginia; Conversion Factors, Volume III, Appendix C.

5. **Answer: C**

A percent may be defined as the identified piece divided by the total.

$$\frac{65 \text{ dry tons}}{70 \text{ dry tons}} \times 100\% = 92.8\%$$

Reference: Water Environment Federation (2008) *Operation of Municipal Wastewater Treatment Plants*, 6th ed.; Manual of Practice No. 11; Water Environment Federation: Alexandria, Virginia; pp 29-16–29-20.

6. **Answer: B**

This problem is best solved in multiple steps

Given

Flow	950	gpm
Total dynamic head	45	ft
Pump efficiency	72	%
Motor efficiency	94	%
Power cost	0.055	$/kW
Hours/day run	3	
Days/week	7	
Weeks/year	52	

First, find the motor horsepower using the equation from the formula sheet at the front of the book.

$$HP_{motor} = \frac{(\text{Flow, gpm})(\text{Head, ft})}{(3960)(\text{Pump Efficiency, \% as a decimal})(\text{Pump Efficiency, \% as a decimal})}$$

$$HP_{motor} = \frac{(950 \text{ gpm})(45 \text{ ft})}{(3960)(0.72)(0.94)}$$

$$HP_{motor} = 15.95$$

Then, convert HP to kW using dimensional analysis.

$$15.95 \text{ HP} \left[\frac{0.746 \text{ kW}}{1 \text{ HP}} \right] = 11.9 \text{ kW}$$

Find total hours of run time.

$$\frac{3 \text{ hours}}{1 \text{ day}} \left[\frac{7 \text{ days}}{1 \text{ week}} \right] \left[\frac{52 \text{ weeks}}{1 \text{ year}} \right] = 1092 \text{ hours of run time per year}$$

Multiply kW by total hours of run time to find number of kWh required.

(11.9 kW)(1092 hours) = 12 994.8 kWh

Finally, use the price of energy as a unit conversion to find the cost to operate this pump for 1 year.

$$12\ 994.8\ \text{kWh}\left[\frac{\$0.055}{1\ \text{kWh}}\right] = \$714.71$$

15.95 HP

0.746 kW

Reference: Water Environment Federation (2008) *Operation of Municipal Wastewater Treatment Plants*, 6th ed.; Manual of Practice No. 11; Water Environment Federation: Alexandria, Virginia; pp 8-15–8-17.

7.   **Answer: C**

A mechanical flowmeter, such as a propeller meter, requires low suspended solids and debris to measure accurately. Solids and debris interfere with the accuracy and efficiency of the flowmeter. Therefore, another type of flowmeter should be used in applications where suspended solids and debris may be present.

Reference: Water Environment Federation (2008) *Operation of Municipal Wastewater Treatment Plants*, 6th ed.; Manual of Practice No. 11; Water Environment Federation: Alexandria, Virginia; p 7-11.

Treatment Process Monitoring, Evaluation, and Adjustment

Class I Job Tasks

- Analyze laboratory data to evaluate and adjust processes;
- Follow industry safety rules and guidelines applicable to treatment processes;
- Implement changes as indicated by laboratory results;
- Operate chemical feed systems (e.g., polymer, ferric);
- Operate SCADA systems;
- Operate the preliminary treatment processes (e.g., screening, grit, flow equalization);
- Operate the primary clarification/sedimentation processes;
- Operate the following secondary treatment processes:
  - Pond/lagoon systems,
  - Secondary clarification/sedimentation processes, and
  - Extended aeration processes (e.g., package, SBR, oxidation ditch); and
- Operate the following disinfection treatment processes:
  - Chlorination processes and
  - Dechlorination processes.

Class I Types of Knowledge Required to Perform Job Tasks

Types of Knowledge	Level of Knowledge
Aeration principles (e.g., mixing, mechanical, diffusers)	Basic
Bacteriological laboratory testing (e.g., coliform, fecal, *E coli*)	Basic
Biological laboratory testing (e.g., BOD, SOUR, CBOD)	Basic
Biosolids disposal and monitoring requirements	Basic
Chemical laboratory testing (e.g., ammonia, phosphorus, alkalinity)	Basic
Chlorinators (e.g., gas, liquid)	Basic
Clarifiers	Basic
Comminuters	Basic
Conveyors	Basic
Dewatering equipment (e.g., centrifuges, presses, drying beds)	Basic
Documentation and recordkeeping	Basic
Effluent disposal and monitoring requirements	Basic
Flow measuring devices (e.g., Parshall flumes, mag meter, venturis)	Basic
Grit removal processes (e.g., grit chamber, aeration, cyclone)	Basic

Types of Knowledge	Level of Knowledge
Hydraulic principles (e.g., mass flow balance, detention time, loading, velocity, HRT)	Basic
Influent monitoring and waste characteristics	Basic
Ozone generation equipment	Basic
Physical laboratory testing (e.g., temperature, solids, dissolved oxygen)	Basic
Pneumatic principles (e.g., troubleshooting actuators, compressors, sprayers)	Basic
Primary treatment processes (e.g., ponds, sedimentation basins)	Basic
Principles of asset management (e.g., preventive, reactive, predictive maintenance)	Basic
Process control instrumentation (e.g., PLCs, SCADA, continuous online monitoring)	Basic
Quality control/quality assurance practices	Basic
Screening technology (e.g., bar screens, microscreens)	Basic
Secondary treatment processes (e.g., activated sludge, MBR, SBR)	Basic
Solids treatment concepts (e.g., dewatering, digestion, thickening)	Basic
Tertiary treatment processes (e.g., media filtration, disinfection, post-aeration, reclaimed recharge)	Basic
Treatment equipment (e.g., pumps, motors, generators)	Basic
Wastewater treatment practices (e.g., sludge age, SRT, MCRT, F:M)	Basic

Class I Sample Questions

1. Preliminary treatment may include

   a) clarifiers, screens, comminution, flow equalization, and pre-aeration.
   b) flow measurement, solids dewatering, flow equalization, and grit removal.
   c) screens, flow measurement, pre-aeration, grit removal, and comminution.
   d) screens, solids disposal, flow measurement, and flow equalization.

2. What is the detention time of a settling tank that is 75 ft (23 m) long, 20 ft (6 m) wide, and 15 ft (4.6 m) deep and has a flow of 2 mgd (7.6 ML/d)?

   a) 2.0 hours
   b) 2.8 hours
   c) 4.1 hours
   d) 1.7 days

3. Which type of foam is caused by the filamentous organism *Nocardia*?

   a) Stiff white foam
   b) Light brown foam

c) Greasy dark-tan foam

d) Very dark black foam

4. When using metal salts for enhanced solids removal, which adjustment may need to occur to maintain a stable process pH value?

a) A form of acid may need to be added to stabilize the process pH.

b) A form of hydroxide may need to be added to stabilize process pH.

c) A form of acid may need to be added to lower the process pH.

d) A form of hydroxide may need to be added to lower the process pH.

5. Ultraviolet disinfection is a form of wastewater disinfection that operates on the principle that radiation

a) alters the reproductive capability of a bacterial cell.

b) oxidizes the bacteria.

c) completely destroys the bacterial cell material.

d) causes the bacteria to coagulate and be removed by sedimentation.

6. Which of the following statements is correct?

a) The coagulation process follows the flocculation process.

b) The flocculation process creates conditions where the collisions of solids are decreased.

c) Coagulation chemicals are traditionally added after flocculation.

d) The flocculation process creates conditions where the collisions of coagulated solids are increased.

7. What is the most common process control parameter monitored for a mixed-media wastewater filtration system?

a) Turbidity

b) pH

c) Suspended solids

d) Alkalinity

8. In thickening or dewatering processes, one of the most common causes of excessive polymer consumption is

a) storage of polymer totes at 20 °C.

b) poor polymer mixing.

c) low total dissolved solids in the makeup water.

d) use of a non-ionic polymer.

9. In anaerobic digestion, effective mixing provides for benefits such as distribution of influent solids throughout the digester for maximum contact with digester microorganisms, reduction of scum buildup, and

    a) creation of tank stratifications.
    b) buildup of solids on the digester floor.
    c) reduction of digester gas production.
    d) dilution of inhibitors, such toxic materials.

10. Operation of an aerobic digester can be improved by which of the following actions?

    a) Feeding directly with unthickened mixed liquor suspended solids
    b) Parallel operation
    c) Utilizing only anoxic operating conditions
    d) Maintaining sufficient dissolved oxygen concentration

11. Nuisance odors from aerobically digested sludge can be reduced by

    a) increasing the organic loading to the digester.
    b) increasing the dissolved oxygen concentration in the digester.
    c) decreasing the available dissolved oxygen.
    d) adding lime to increase the pH.

12. A digester has a diameter of 100 ft and the cylindrical portion is 25 ft tall. It has a bottom cone with a depth of 15 ft. The liquid level in the digester is at 20 ft from the bottom of the cylinder. What is the volume of sludge in the digester in cubic feet?

    a) 157 000 cu ft
    b) 166 200 cu ft
    c) 196 250 cu ft
    d) 225 000 cu ft

13. Which of the following is characteristic of fresh raw municipal wastewater?

    a) Strong acidic odor
    b) Gray color
    c) Low temperature compared to ambient temperature
    d) Minimum daily flow occurring in the afternoon

14. A significant increase in wastewater temperature for a short period of time typically indicates the presence of

    a) stormwater inflow.
    b) groundwater infiltration.

c) extremely high ambient temperature.

d) an industrial discharge.

15. Wastewater flows typically vary consistently during days, weeks, seasons, and years. The daily (diurnal) flow variation depends largely on

a) inflow and infiltration.

b) size and configuration of the collection system.

c) operation of influent pumping stations.

d) hours of operation for schools.

16. The sludge on a dewatering sand bed is dusty and crumbles. The sludge

a) cake is ready for removal from the bed.

b) is excessively dry.

c) requires more drying time.

d) is a breeding site for flies—add polymer.

Answers

1. **Answer: C**

   Preliminary treatment of wastewater includes screening, grit removal, odor control (where appropriate), and flow measurement.

   Reference: Water Environment Federation (2008) *Operation of Municipal Wastewater Treatment Plants*, 6th ed.; Manual of Practice No. 11; Water Environment Federation: Alexandria: Virginia; p 18-2.

2. **Answer: A**

   The formula for detention time can be found in the formula sheet in the front of this manual.

   $$\text{Detention Time} = \frac{\text{Volume}}{\text{Flow}}$$

   The formula requires volume, which is not given; however, information to find volume is given.

   Volume = (Length)(Width)(Height)
   Volume = (75 ft)(20 ft)(15 ft)
   Volume = 22 500 cu ft

For the formula to work, the units for volume and flow must be compatible. Because flow is given in million gallons per day, the volume must be in million gallons.

$$22\,500 \text{ cu ft} \left[\frac{7.48 \text{ gal}}{1 \text{ cu ft}} \right] \left[\frac{1 \text{ MG}}{1\,000\,000 \text{ gal}} \right] = 0.168 \text{ mil. gal}$$

Now that volume is in the correct units, use the detention time formula:

$$\text{Detention Time} = \frac{\text{Volume}}{\text{Flow}}$$

$$\text{Detention Time} = \frac{0.168 \text{ mil. gal}}{2 \text{ mgd}}$$

Detention Time = 0.084 days

Convert days to hours:

$$0.084 \text{ days} \left[\frac{24 \text{ hours}}{1 \text{ day}} \right] = 2.0 \text{ hours}$$

Reference: California State University, Sacramento (2008) *Operation of Wastewater Treatment Plants,* 7th ed.; California State University: Sacramento, California; Volume I, p 135.

3. **Answer: C**

Water resource recovery facilities with a significant quantity of the filamentous organism *Nocardia* will have a thick greasy dark-tan foam that will get trapped wherever overflow outlet structures are not present. Spray systems (with or without chlorine) are often used to help manage foam issues. A more effective method for *Nocardia* control is to physically remove the *Nocardial* foam from the reactor. Because the filaments concentrate in the foam, *Nocardia* will be selectively removed from the system by removing foam. Sidestreams containing *Nocardia* foam and removed foam must be disposed of and not allowed to recycle back to the plant.

Reference: Water Environment Federation (2008) *Operation of Municipal Wastewater Treatment Plants,* 6th ed.; Manual of Practice No. 11; Water Environment Federation: Alexandria, Virginia; pp 20-48, 20-115.

4. **Answer: B**

Metal salts typically will consume alkalinity and lower process pH if sufficient alkalinity is not present in the source water. A buffer chemical would provide alkalinity to stabilize the process pH.

Reference: Water Environment Federation (2008) *Operation of Municipal Wastewater Treatment Plants,* 6th ed.; Manual of Practice No. 11; Water Environment Federation: Alexandria: Virginia; p 24-66.

5. **Answer:** A

Ultraviolet disinfection of bacterial cells alters the genetic material (DNA and RNA) of the bacterial cell, preventing the reproduction of that cell.

Reference: Water Environment Federation (2008) *Operation of Municipal Wastewater Treatment Plants*, 6th ed.; Manual of Practice No. 11; Water Environment Federation: Alexandria, Virginia; p 26-7.

6. **Answer:** D

Traditionally, the flocculation process is preceded by chemical coagulation. The flocculation process is designed to increase coagulated solids collisions. Coagulation chemicals need to be added before flocculation.

Reference: Water Environment Federation (2008) *Operation of Municipal Wastewater Treatment Plants*, 6th ed.; Manual of Practice No. 11; Water Environment Federation: Alexandria, Virginia; pp 24-14–24-15.

7. **Answer:** A

pH is not a common parameter to be monitored for a mixed-media wastewater filtration system. Suspended solids represent a common process control quantifier, although they are not as quickly determined as turbidity. Alkalinity is not a performance quantifier.

Reference: Water Environment Federation (2008) *Operation of Municipal Wastewater Treatment Plants*, 6th ed.; Manual of Practice No. 11; Water Environment Federation: Alexandria, Virginia; p 24-45.

8. **Answer:** B

The type of polymer preparation and feed system is determined by the type of polymer product used because different mixing systems are required to effectively mix the different types of polymer. The method of mixing the sludge and polymer is important because efficient mixing ensures the most consistent polymer performance.

Reference: Water Environment Federation (1997) *Basic Maintenance of Belt Filter Presses*; Water Environment Federation: Alexandria, Virginia; p 24.

9. **Answer:** D

Reference: Water Environment Federation (2008) *Operation of Municipal Wastewater Treatment Plants*, 6th ed.; Manual of Practice No. 11; Water Environment Federation: Alexandria, Virginia; pp 30-17–30-18.

10. **Answer:** D

Maintaining adequate oxygen levels allows the biological process to take place and prevents objectionable odors. In aerobic digesters, a typical dissolved oxygen concentration range is from 0.1 to 3.0 mg/L. Inadequate oxygen levels result in incomplete digestion.

Reference: Water Environment Federation (2008) *Operation of Municipal Wastewater Treatment Plants*, 6th ed.; Manual of Practice No. 11; Water Environment Federation: Alexandria, Virginia; p 31-27.

11. Answer: B

Inadequate dissolved oxygen levels result in incomplete digestion and odor problems.

Reference: Water Environment Federation (2008) *Operation of Municipal Wastewater Treatment Plants*, 6th ed.; Manual of Practice No. 11; Water Environment Federation: Alexandria, Virginia; p 31-27.

12. Answer: C

First, find the volume of the upper portion of the tank, which is a cylinder.

The digester has two components: the cylinder portion on top and the cone on the bottom.

Volume of a Cylinder = (0.785)(Diameter²)(Height)
Volume of a Cylinder = (0.785)(100 ft)(100 ft)(20 ft)
Volume of a Cylinder = 157 000 cu ft

Volume of a Cone = 1/3(0.785)(Diameter²)(Height)
Volume of a Cone = 1/3(0.785)(100 ft)(100 ft)(15 ft)
Volume of a Cone = 39 250 cu ft

Add the two volumes together to get the total volume.

157 000 cu ft + 39 250 cu ft = 196 250 cu ft

Reference: Water Environment Federation (2008) *Operation of Municipal Wastewater Treatment Plants*, 6th ed.; Manual of Practice No. 11; Water Environment Federation: Alexandria, Virginia; p 31-62.

13. Answer: B

Typical domestic (municipal) raw wastewater often appears gray. The color can vary significantly, including appearing black when wastewater goes septic to brightly colored as a result of industrial discharges.

Reference: Water Environment Federation (2008) *Operation of Municipal Wastewater Treatment Plants,* 6th ed.; Manual of Practice No. 11; Water Environment Federation: Alexandria, Virginia; p 17-5.

14. **Answer:** D

A significant increase in temperature for a short period of time typically indicates the presence of a high-temperature industrial discharge.

Reference: Water Environment Federation (2008) *Operation of Municipal Wastewater Treatment Plants,* 6th ed.; Manual of Practice No. 11; Water Environment Federation: Alexandria, Virginia; p 17-5.

15. **Answer:** B

In general, the smaller the collection system, the greater the diurnal variation. For small systems, the peak hour flow may be up to five times higher than the average daily flow. For larger systems, the diurnal flow variation tends to be much less because wastewater travel time is greater. Daily flows for water resource recovery facilities tend to peak between 8:00 a.m. and 10:00 a.m. and again between 4:00 p.m. and 7:00 p.m. Minimum daily flows typically occur late in the evening and early morning.

Reference: Water Environment Federation (2008) *Operation of Municipal Wastewater Treatment Plants,* 6th ed.; Manual of Practice No. 11; Water Environment Federation: Alexandria, Virginia; p 17-4.

16. **Answer:** B

If the sludge is dusty, it is too dry.

Reference: Water Environment Federation (2008) *Operation of Municipal Wastewater Treatment Plants,* 6th ed.; Manual of Practice No. 11; Water Environment Federation: Alexandria, Virginia; p 33-53.

Class II Job Tasks

- Add chemicals to disinfect and deodorize water and other liquids (e.g., ammonia, chlorine, lime);
- Analyze laboratory data to evaluate and adjust processes;
- Follow industry safety rules and guidelines applicable to treatment processes;
- Implement changes as indicated by laboratory results;
- Operate chemical feed systems (e.g., polymer, ferric);
- Operate odor control systems (e.g., biofilters, scrubbers);
- Operate SCADA systems;
- Operate the preliminary treatment processes (e.g., screening, grit, flow equalization);
- Operate the primary clarification/sedimentation processes;
- Operate the following secondary treatment processes:
  - Attached growth/fixed film processes (e.g., RBC, trickling filter),
  - Secondary clarification/sedimentation processes,

- - Extended aeration processes (e.g., package, SBR, oxidation ditch), and
  - Conventional activated sludge processes (e.g., step feed, plug flow, complete mix, MBR);
- Operate the nutrient removal systems;
- Operate the following disinfection treatment processes:
  - Chlorination processes,
  - Dechlorination processes, and
  - Disinfection processes (e.g., UV, ozone); and
- Operate the following solids treatment processes:
  - Aerobic digestion process and
  - Mechanical dewatering processes (e.g., presses, centrifuges).

Class II Types of Knowledge Required to Perform Job Tasks

Types of Knowledge	Level of Knowledge
Aeration principles (e.g., mixing, mechanical, diffusers)	Basic
Bacteriological laboratory testing (e.g., coliform, fecal, *E coli*)	Basic
Biological laboratory testing (e.g., BOD, SOUR, CBOD)	Basic
Biosolids disposal and monitoring requirements	Basic
Chemical laboratory testing (e.g., ammonia, phosphorus, alkalinity)	Basic
Chlorinators (e.g., gas, liquid)	Basic
Clarifiers	Basic
Comminuters	Basic
Conveyors	Basic
Dewatering equipment (e.g., centrifuges, presses, drying beds)	Basic
Documentation and recordkeeping	Basic
Effluent disposal and monitoring requirements	Intermediate
Flow measuring devices (e.g., Parshall flumes, mag meter, venturis)	Basic
Grit removal processes (e.g., grit chamber, aeration, cyclone)	Basic
Hydraulic principles (e.g., mass flow balance, detention time, loading, velocity, HRT)	Basic
Influent monitoring and waste characteristics	Basic
Ozone generation equipment	Basic
Physical laboratory testing (e.g., temperature, solids, dissolved oxygen)	Basic
Pneumatic principles (e.g., troubleshooting actuators, compressors, sprayers)	Basic
Primary treatment processes (e.g., ponds, sedimentation basins)	Intermediate
Principles of asset management (e.g., preventive, reactive, predictive maintenance)	Basic

Types of Knowledge	Level of Knowledge
Process control instrumentation (e.g., PLCs, SCADA, continuous online monitoring)	Basic
Quality control/quality assurance practices	Basic
Screening technology (e.g., bar screens, microscreens)	Basic
Secondary treatment processes (e.g., activated sludge, MBR, SBR)	Intermediate
Solids treatment concepts (e.g., dewatering, digestion, thickening)	Basic
Tertiary treatment processes (e.g., media filtration, disinfection, post-aeration, reclaimed recharge)	Basic
Treatment equipment (e.g., pumps, motors, generators)	Basic
Wastewater treatment practices (e.g., sludge age, SRT, MCRT, F:M)	Basic

Class II Sample Questions

1. The total suspended solids concentration from an activated sludge secondary clarifier is increasing. The sludge volume index test is within normal range. What should be checked first?

    a) Waste activated sludge pumping rate
    b) Secondary effluent biochemical oxygen demand concentration
    c) Secondary clarifier blanket depth
    d) Heavy metals concentrations in the raw influent wastewater

2. Where in a facultative lagoon are anaerobic bacteria located?

    a) At the end, nearest the outlet
    b) Anaerobic bacteria are not in facultative lagoons
    c) At the inlet surface
    d) Near the bottom in the sludge blanket

3. If the biomass on a rotating biological contactor turns white, what might be the cause?

    a) Biochemical oxygen underload
    b) Low dissolved oxygen
    c) Sulfur-loving bacteria
    d) Hydraulic overload

4. Common coagulation chemicals used in conjunction with a tertiary filter to assist with phosphorus removal include

    a) methanol and sodium hydroxide.
    b) ferric chloride and aluminum sulfate.

c) sodium hypochlorite and sodium bisulfite.

d) activated carbon and diatomaceous earth.

5. A biotower distributor arm is designed to apply wastewater to a biotower media at a rate of 2784 gal (10.56 m³) per revolution but currently the biotower is receiving a total flowrate of 1 mgd (3785 m³/d). If the speed of the distributor is maintained at a constant rate, what is the speed of the distributor?

a) 0.25 rpm

b) 2.5 rpm

c) 3 rpm

d) 4 rpm

6. Chlorination tends to be more effective when wastewater has a lower pH. This is primarily attributable to

a) a higher percentage of hypochlorite ion (OCl^-) at a lower pH.

b) a higher percentage of hydrochlorous acid (HOCl) at a lower pH.

c) coliform bacteria growth inhibition at a lower pH.

d) the achievement of proper mixing at a lower pH.

7. The following are interferences to UV transmittance EXCEPT

a) color.

b) turbidity.

c) organics.

d) pH.

8. During the lime stabilization process, lime is added to

a) maintain a pH of 6.5 to 8.5 for 8 hours.

b) prevent the release of ammonia odors.

c) maintain a pH of 12 for a minimum of 2 hours.

d) turn the biosolids to a greyish white color.

9. What gas is produced in the anaerobic digester and can be used as fuel?

a) propane

b) methane

c) oxygen

d) carbon dioxide

10. Which class(es) of biosolids must meet vector attraction requirements before being land applied?

   a) Neither Class A or Class B
   b) Class A
   c) Class B
   d) Class A and Class B

11. Given the following information, find the solids loading rate (SLR) for a secondary clarifier:

   - Influent flow = 2.0 mgd
   - Return activated sludge flow = 70% of influent flow
   - Mixed liquor suspended solids (MLSS) concentration = 3150 mg/L
   - Return activated sludge/waste activated sludge concentration = 7650 mg/L
   - Clarifier diameter = 65 ft
   - Clarifier depth = 12 ft

   a) 24 lb/d/sq ft (118 kg/m²·d)
   b) 27 lb/d/sq ft (132 kg/m²·d)
   c) 55 lb/d/sq ft (320 kg/m²·d)
   d) 59 lb/d/sq ft (287 kg/m²·d)

12. Which of the following is **NOT** a common effluent discharge reuse application?

   a) Landscape irrigation
   b) Boiler makeup
   c) Drinking water system intake water
   d) Aquifer recharge

13. When septic conditions occur in the collection system, what physical characteristics of raw wastewater will be observed?

   a) Low suspended solids concentration and methane production
   b) Dark color and hydrogen sulfide odor
   c) Higher temperature and low flowrate
   d) High turbidity and foams produced

14. If possible, what should be recirculated through sludge lines on a weekly basis to reduce grease buildup?

   a) Chemicals
   b) Digested sludge
   c) Effluent
   d) Raw wastewater

Answers

1. **Answer: C**

    Blanket depth should be measured at least one to three times per day and more often during high flows. Typically, clarifiers are operated so that the solids blanket depth is between 0.5 and 3.0 ft (0.15 and 0.9 m) up from the bottom. A blanket depth allowed to increase too high could cause solids to be carried over the clarifier weirs, resulting in effluent compliance violations.

    Reference: Water Environment Federation (2008) *Operation of Municipal Wastewater Treatment Plants*, 6th ed.; Manual of Practice No. 11; Water Environment Federation: Alexandria, Virginia; p 20-44.

2. **Answer: D**

    As wastewater enters a lagoon, the heavy solids settle out near the inlet bottom, where anaerobic bacteria stabilize the organic matter.

    Reference: Water Environment Federation (2008) *Operation of Municipal Wastewater Treatment Plants*, 6th ed.; Manual of Practice No. 11; Water Environment Federation: Alexandria, Virginia; p 23-4.

3. **Answer: C**

    The biomass on a rotating biological contactor should be "shaggy looking" and brownish-gray in color. If the biomass turns white, it means the sulfur content of the influent is high, and sulfur-loving bacteria are flourishing.

    Reference: California State University, Sacramento (2008) *Operation of Wastewater Treatment Plants*, 7th ed.; California State University: Sacramento, California; Volume I, p 228.

4. **Answer: B**

    Reference: Water Environment Federation (2008) *Operation of Municipal Wastewater Treatment Plants*, 6th ed.; Manual of Practice No. 11; Water Environment Federation: Alexandria, Virginia; p 24-27.

5. **Answer: A**

    Use dimensional analysis for the solution, as follows:

    $$1 \quad \rule{2cm}{0.4pt} \quad \rule{2cm}{0.4pt} \quad \rule{2cm}{0.4pt} \quad \rule{2cm}{0.4pt}$$

    U.S. Customary units:

    $$\frac{1\,000\,000 \text{ gal}}{d}\left[\frac{1 \text{ rev}}{2784 \text{ gal}}\right]\left[\frac{1 \text{ d}}{24 \text{ h}}\right]\left[\frac{1 \text{ h}}{60 \text{ min}}\right] = 0.25 \text{ rev/min}$$

SI units:

$$\frac{3785 \text{ m}^3}{\text{d}} \left[\frac{1 \text{ rev}}{10.56 \text{ m}^3}\right]\left[\frac{1 \text{ d}}{24 \text{ h}}\right]\left[\frac{1 \text{ h}}{60 \text{ min}}\right] = 0.25 \text{ rev/min}$$

Reference: Water Environment Federation (2008) *Operation of Municipal Wastewater Treatment Plants*, 6th ed.; Manual of Practice No. 11; Water Environment Federation: Alexandria, Virginia; pp 21-14–21-15 (note: reference is for visualization of the process only).

6.   **Answer: B**

Hydrochlorous acid is a better disinfectant than OCl^-. The ratio of $HOCl$ to OCl^- increases as the pH is reduced.

Reference: Water Environment Federation; American Society of Civil Engineers; Environmental & Water Resources Institute (2010) *Design of Municipal Wastewater Treatment Plants*, 5th ed.; WEF Manual of Practice No. 8; ASCE Manuals and Reports on Engineering Practice No. 76; Water Environment Federation: Alexandria, Virginia; pp 19-50–19-51.

7.   **Answer: D**

UV transmittance interferences include color, suspended solids, turbidity, iron, organics, chlorine, and high levels of nitrates.

Reference: Water Environment Federation (2008) *Operation of Municipal Wastewater Treatment Plants*, 6th ed.; Manual of Practice No. 11; Water Environment Federation: Alexandria, Virginia; p 26-14.

8.   **Answer: C**

Maintaining a pH of 12 for a minimum of 2 hours typically destroys or inhibits pathogens and the microorganisms involved in the decomposition of the sludge.

Reference: Water Environment Federation (2008) *Operation of Municipal Wastewater Treatment Plants*, 6th ed.; Manual of Practice No. 11; Water Environment Federation: Alexandria, Virginia; p 32-16.

9.   **Answer: B**

Digester gas is composed primarily of methane (60 to 65%) and carbon dioxide (35 to 40%). The gas has a lower heating value of 22 400 kJ/m^3 (600 Btu/cu ft) because it typically contains only 60 to 65% methane. It is a valuable resource that can be used to meet a water resource recover facility's energy requirements.

Reference: Water Environment Federation (2008) *Operation of Municipal Wastewater Treatment Plants*, 6th ed.; Manual of Practice No. 11; Water Environment Federation: Alexandria, Virginia; p 30-37.

10. **Answer:** D

Both Class A and Class B biosolids must meet vector attraction reduction requirements. These requirements are intended to reduce the putrescibility of the solids and make them less attractive as a potential food source to birds, insects, and burrowing animals that could potentially transport biosolids away from the application site.

Reference: Water Environment Federation (2008) *Operation of Municipal Wastewater Treatment Plants*, 6th ed.; Manual of Practice No. 11; Water Environment Federation: Alexandria, Virginia; p 30-10.

11. **Answer:** B

Typical solids loading rates (SLRs) in the safe operating range are 171 kg/m²·d (35 lb/d/sq ft) or less. The maximum SLR at specific sludge volume indexes (SVIs) will differ for different water resource recovery facilities. It is best to determine the actual SLR limits for your facility. The poorer the activated sludge settleability, as measured by the SVI, however, the lower the SLR should be.

First, find the clarifier surface area in square feet:

Area = (0.785)(diameter²)
Area = (0.785)(65 ft)(65 ft)
Area = 3316.6 sq ft

Then, find the SLR:

$$\text{Solids Loading Rate, lb/d/sq ft} = \frac{(\text{Flow}_{influent} + \text{Flow}_{RAS})(\text{MLSS, mg/L})(8.34\ \text{lb/mil} \cdot \text{mil. gal})}{\text{Surface Area, sq ft}}$$

$$\text{Solids Loading Rate, lb/d/sq ft} = \frac{[(2.0\ \text{mgd} + (2.0\ \text{mgd})(0.7)](3150\ \text{mg/L})(8.34\ \text{lb/mil. gal})}{3316.6\ \text{sq ft}}$$

$$\text{Solids Loading Rate, lb/d/sq ft} = \frac{(3.4\ \text{mgd})(3150\ \text{mg/L})(8.34\ \text{lb/mil. gal})}{3316.6\ \text{sq ft}}$$

$$\text{Solids Loading Rate, lb/d/sq ft} = 26.9$$

Reference: Water Environment Federation (2008) *Operation of Municipal Wastewater Treatment Plants*, 6th ed.; Manual of Practice No. 11; Water Environment Federation: Alexandria, Virginia; p 20-44.

12. **Answer:** C

Reference: Water Environment Federation (2012) *Membrane Bioreactors*; Manual of Practice No. 36; Water Environment Federation: Alexandria, Virginia; p 9.

13. **Answer:** B

Septic influent wastewater will appear dark in color and have a "rotten egg" odor because of the presence of hydrogen sulfide.

Reference: Water Environment Federation (2008) *Operation of Municipal Wastewater Treatment Plants,* 6th ed.; Manual of Practice No. 11; Water Environment Federation, Alexandria, Virginia; pp 17-5–17-6.

14. **Answer:** B

If possible, warm anaerobically digested sludge should be recirculated through the line for an hour each week if grease tends to build up on pipe walls.

One means of combatting grease accumulation requires filling underflow lines with digested sludge and allowing the sludge to remain undisturbed for at least 1 day before removing the digested sludge and dislodged grease.

Reference: Water Environment Federation (2008) *Operation of Municipal Wastewater Treatment Plants,* 6th ed.; Manual of Practice No. 11; Water Environment Federation: Alexandria, Virginia; p 29-19.

Class III Job Tasks

- Add chemicals to disinfect and deodorize water and other liquids (e.g., ammonia, chlorine, lime);
- Analyze laboratory data to evaluate and adjust processes;
- Follow industry safety rules and guidelines applicable to treatment processes;
- Implement changes as indicated by laboratory results;
- Operate chemical feed systems (e.g., polymer, ferric);
- Operate SCADA systems;
- Operate the preliminary treatment processes (e.g., screening, grit, flow equalization);
- Operate the primary clarification/sedimentation processes;
- Operate the following secondary treatment processes:
  - Secondary clarification/sedimentation processes,
  - Extended aeration processes (e.g., package, SBR, oxidation ditch), and
  - Conventional activated sludge processes (e.g., step feed, plug flow, complete mix, MBR);
- Operate the nutrient removal systems;
- Operate the following disinfection treatment processes:
  - Chlorination processes,
  - Dechlorination processes, and

- ○ Disinfection processes (e.g., UV, ozone); and
- Operate the following solids treatment processes:
  - ○ Anaerobic digestion process,
  - ○ Mechanical dewatering processes (e.g., presses, centrifuges), and
  - ○ Solids thickening processes (e.g., DAF, belt, rotary drum).

Class III Types of Knowledge Required to Perform Job Tasks

Types of Knowledge	Level of Knowledge
Aeration principles (e.g., mixing, mechanical, diffusers)	Intermediate
Bacteriological laboratory testing (e.g., coliform, fecal, *E coli*)	Intermediate
Biological laboratory testing (e.g., BOD, SOUR, CBOD)	Intermediate
Biosolids disposal and monitoring requirements	Intermediate
Chemical laboratory testing (e.g., ammonia, phosphorus, alkalinity)	Intermediate
Chlorinators (e.g., gas, liquid)	Intermediate
Clarifiers	Intermediate
Comminuters	Intermediate
Conveyors	Intermediate
Dewatering equipment (e.g., centrifuges, presses, drying beds)	Intermediate
Documentation and recordkeeping	Intermediate
Effluent disposal and monitoring requirements	Intermediate
Flow measuring devices (e.g., Parshall flumes, mag meter, venturis)	Intermediate
Grit removal processes (e.g., grit chamber, aeration, cyclone)	Intermediate
Hydraulic principles (e.g., mass flow balance, detention time, loading, velocity, HRT)	Intermediate
Influent monitoring and waste characteristics	Intermediate
Ozone generation equipment	Intermediate
Physical laboratory testing (e.g., temperature, solids, dissolved oxygen)	Intermediate
Pneumatic principles (e.g., troubleshooting actuators, compressors, sprayers)	Intermediate
Primary treatment processes (e.g., ponds, sedimentation basins)	Advanced
Principles of asset management (e.g., preventive, reactive, predictive maintenance)	Intermediate
Process control instrumentation (e.g., PLCs, SCADA, continuous online monitoring)	Intermediate
Quality control/quality assurance practices	Intermediate
Screening technology (e.g., bar screens, microscreens)	Intermediate
Secondary treatment processes (e.g., activated sludge, MBR, SBR)	Intermediate
Solids treatment concepts (e.g., dewatering, digestion, thickening)	Intermediate

Types of Knowledge	Level of Knowledge
Tertiary treatment processes (e.g., media filtration, disinfection, post-aeration, reclaimed recharge)	Intermediate
Treatment equipment (e.g., pumps, motors, generators)	Intermediate
Wastewater treatment practices (e.g., sludge age, SRT, MCRT, F:M)	Intermediate

Class III Sample Questions

1. Solids loss from an activated sludge secondary clarifier occurs when more solids are entering the secondary clarifier than can be settled and then removed in the return activated sludge (RAS) flow. Often, this overloaded condition can be relieved if the mixed liquor suspended solids (MLSS) concentration and/or the solids loading rate to the secondary clarifier can be reduced. Which of the following correctly lists practical steps an operator might take to provide this relief and avoid a potential permit violation?

    a) Put additional secondary clarifiers on-line, put additional aeration basins on-line, and step-feed more primary clarifier effluent flow to the end of the aeration basin

    b) Bypass primary effluent flow around secondary treatment, divert MLSS flow to the digesters, and turn off the air in the aeration basins

    c) Take secondary clarifiers off line, take aeration basins off line, and increase the air flow to the aeration basins

    d) Increase wasting, increase the food-to-microorganism ratio, and add polymer to the MLSS

2. How is the target mean cell residence time (MCRT) maintained in the activated sludge system?

    a) Return activated sludge rates

    b) Food-to-microorganism ratio

    c) Sludge volume index

    d) Solids wasting

3. The sludge volume index is 210 mL/g. The microscopic examination shows filamentous organisms extending beyond the floc. What should be the next step?

    a) Decrease the mean cell residence time

    b) Increase the wasting rate

    c) Identify the filamentous organisms

    d) Use lime and raise the pH above 8.5

4. A light chocolate-colored foam has begun to appear in the sequencing batch reactor. A microscopic exam shows *Nocardia* as being the reason for the foam. What is an effective way to remove it?

    a) Increase the mean cell residence time
    b) Physically remove it and send it to a landfill
    c) Add cationic polymer to the waste
    d) Feed chlorine to the return activated sludge

5. Denitrification can be achieved in an oxidation ditch using all of the following EXCEPT

    a) increasing the recycle rate.
    b) turning off one or more rotors.
    c) turning the aerators off at least twice a day.
    d) creating zones with and without aeration.

6. The biological nutrient removal facility must fully nitrify and has a nitrate goal of 5 mg/L as nitrogen. If the effluent nitrate is 8 mg/L as nitrogen, the ammonia concentration is non-detectable, return activated sludge (RAS) is 35% of influent flow, and the internal mixed liquor recycle rate (IMLR) rate is 200% of influent flow. What change should the operator make to reduce the effluent nitrate concentration further?

    a) RAS pump has failed; there is insufficient mixed liquor suspended solids to meet the incoming nitrogen load
    b) A toxic load has entered the facility; increase the dissolved oxygen setpoint to 3.0 mg/L
    c) The IMLR rate is too low; increase IMLR rate
    d) Waste sludge operation not stabilized; stop wasting

7. Determine the volumetric chemical feed rate in milliliters/minute to inject 1500 lb/d (682 kg/d) of sodium hydroxide to a return activated sludge line at a constant rate. Sodium hydroxide is delivered at 50% solution and has a specific gravity of 1.53. Solutions are distributed by percent (%) mass and there are 3.785 L/gal.

    a) 79 mL/min
    b) 160 mL/min
    c) 620 mL/min
    d) 945 mL/min

8. Lime coagulation will be most successful in removing which of the following heavy metals?

    a) Mercury
    b) Arsenic

    c)   Cadmium

    d)   Selenium

9.   Which of the following is an advantage of incineration?

    a)   Good as an interim or emergency stabilization method

    b)   Total pathogen destruction

    c)   Low total capital cost

    d)   Requires an auxiliary fuel source

10.   The local water resource recovery facility has three 60-ft (18-m) diameter anaerobic digesters that are 35 ft deep. Given the following information:

Thickened sludge flow = 11 L/s (250 000 gpd)
- Thickened sludge percent solids = 4.1% and
- Thickened sludge percent volatile solids = 78%.

Digested sludge transferred per day = 6.3 L/s (144 000 gpd)
- Digested sludge percent solids = 2.61% and
- Digested sludge percent volatile solids = 64%.

Calculate the volatile solids reduction.

    a)   14%

    b)   18%

    c)   50%

    d)   55%

11.   What type of microorganism is prevalent in the second stage of anaerobic digestion?

    a)   Methane-forming bacteria

    b)   Fungi

    c)   Aerobic bacteria

    d)   Acid-forming bacteria

12.   Interpret the results for a sequencing batch reactor facility designed to remove both nitrogen and phosphorus. Today's results show that effluent total phosphorus is 2 mg/L as phosphorus, ammonia nitrogen is 8 mg/L, nitrate nitrogen < 5 mg/L, and biochemical oxygen demand (BOD) < 25 mg/L. What adjustment should the operator make to decrease the effluent ammonia concentration?

    a)   Increase the percentage of air-on cycle time

    b)   Increase solids residence time to decrease effluent BOD

    c) Increase the anoxic react time

    d) Ensure that the dissolved oxygen concentration is no more than 0.5 mg/L throughout the entire react phase

13. The primary advantages of a membrane bioreactor (MBR) over conventional activated sludge facilities are

    a) low energy use and the ability to attenuate large increases in hydraulic loading.

    b) a reduced mixed liquor recycle rate and a lower overall energy use.

    c) membranes are not affected by poorly settling activated sludge and have a lower overall process footprint.

    d) lower odor and the ability to remove nitrogen.

14. During a sludge settling test, you observe the presence of clear supernatant above poorly settling sludge. This is likely caused by what?

    a) Very low food-to-microorganism ratio

    b) Excessive amounts of filamentous organisms in the sludge

    c) High nutrient levels

    d) Too high influent biochemical oxygen demand concentration

15. Which of the following might occur in anaerobic digesters that are organically overloaded?

    a) Rotten egg odor

    b) Temperature increase

    c) pH increase

    d) Excessive foaming

16. A water resource recovery facility receives an average daily flow of 12.5 mgd. The influent BOD concentration is 320 mg/L. If the primary clarifier removes 35% of the influent BOD, what is the organic load to the secondary treatment process in pounds per day?

    a) 11 676 lb/d

    b) 16 446 lb/d

    c) 21 684 lb/d

    d) 33 360 lb/d

17. What will happen in an anaerobic digester when the temperature inside the tank is measured at 10 °C (50 °F)?

    a) Digestion almost ceases.

    b) Methane production increases.

    c) *Nocardia* foams occur.

    d) Temperature has no effect on anaerobic digestion.

Answers

1. **Answer:** A

    Clarifier overloading occurs when solids enter the clarifier faster than they can be settled. Solids overloading is related to the facility flow, return flow, MLSS concentration, and clarifier surface area. Determine the solids loading rate to the clarifier according to the following formula resulting in kilograms per square meters per hour in SI units and pounds per hour per square feet in U.S. customary units:

    $$\text{Solids Loading Rate} = \frac{[(\text{Influent Flow} + \text{RAS Flow})(\text{MLSS Concentration})(\text{Conversion factor})]}{(24\ \text{hr/d})(\text{Clarifier surface area})}$$

    The solids loading rate may be reduced by reducing RAS flow, reducing MLSS concentration, or by increasing the amount of clarifier surface area. The MLSS concentration may be reduced by placing additional aeration basins into service or by using step-feed to dilute the MLSS before it enters the clarifiers.

    Reference: Water Environment Federation (2008) *Operation of Municipal Wastewater Treatment Plants*, 6th ed.; Manual of Practice No. 11; Water Environment Federation: Alexandria, Virginia; p 20-120.

2. **Answer:** D

    The MCRT is defined as the number of days a microorganism stays in the activated sludge system before it is wasted out of the system. A longer MCRT leads to more solids being kept in the system and an increase in MLSS concentration. The MCRT is calculated in days. The basic equation for MCRT is the pounds of MLSS in the system divided by the pounds leaving the system. MCRT is controlled by adjusting the wasting rate, which is pounds per day leaving the system.

    Reference: Water Environment Federation (2008) *Operation of Municipal Wastewater Treatment Plants*, 6th ed.; Manual of Practice No. 11; Water Environment Federation: Alexandria, Virginia; p 20-40.

3. **Answer:** C

    It is important to properly identify filamentous organisms. Different filaments thrive under different environmental conditions, including low dissolved oxygen, excessive mean cell residence time, nutrient deficiency, the presence of septic conditions, industrial wastewater, and other reasons.

    Reference: Water Environment Federation (2008) *Operation of Municipal Wastewater Treatment Plants*, 6th ed.; Manual of Practice No. 11; Water Environment Federation: Alexandria, Virginia; p 20-46.

4.   **Answer:** B

Physical removal and disposal to a landfill is the best way to rid a plant of *Nocardia*. Using chlorine or lime can upset the whole process and cause more problems.

Reference: Water Environment Federation (2008) *Operation of Municipal Wastewater Treatment Plants*, 6th ed.; Manual of Practice No. 11; Water Environment Federation: Alexandria, Virginia; p 20-48.

5.   **Answer:** A

Denitrification can be achieved by turning off one or more aeration rotors to create an anoxic zone, cycling the aeration by turning the aerators off at least twice a day, or using phased ditch operation.

Reference: Water Environment Federation (2008) *Operation of Municipal Wastewater Treatment Plants*, 6th ed.; Manual of Practice No. 11; Water Environment Federation: Alexandria, Virginia, p 22-34.

6.   **Answer:** C

The IMLR pumping rates may be 100 to 400% of the influent flow, depending on such factors as the target effluent nitrate concentration. Increasing IMLR rate allows more nitrate to be returned to the anoxic zone. Provided sufficient biochemical oxygen demand (BOD) is available, increasing the IMLR will reduce effluent nitrate concentrations. Denitrification requires 4 mg/L of BOD for every 1 mg/L of nitrate-nitrogen converted to nitrogen gas.

Reference: Water Environment Federation (2008) *Operation of Municipal Wastewater Treatment Plants*, 6th ed.; Manual of Practice No. 11; Water Environment Federation: Alexandria, Virginia; p 22-28, Figure 22.11.

7.   **Answer:** C

It is important to understand that 50% solution equals 50 lb (23 kg) of chemical (NaOH) per 100 lb (45 kg) of solution. The per-gallon weight of solution is calculated by multiplying the weight of 1 gal of water by the specific gravity of the solution. Use dimensional analysis, as follows:

$$\frac{1500 \text{ lb NaOH}}{d} \times \frac{100 \text{ lb soln}}{50 \text{ lb NaOH}} \times \frac{3785 \text{ mL}}{\text{gal}} \times \frac{1 \text{ gal solution}}{8.34 \times 1.53 \text{ lb solution}} \times \frac{1 \text{ d}}{1440 \text{ min}} = 617.96 \frac{\text{mL}}{\text{min}}$$

Round to 620 mL/min.

$$\text{International System of Units: } \frac{682 \text{ kg NaOH}}{d} \times \frac{100 \text{ kg solution}}{50 \text{ kg NaOH}} \times \frac{1000 \text{ mL}}{1 \text{ L}}$$

$$\times \frac{1 \text{ L}}{1 \times 1.53 \text{ kg solution}} \times \frac{1 \text{ d}}{1440 \text{ min}} = 619.1 \frac{\text{mL}}{\text{min}}$$

Distractors:

a) Assuming 50 lb/gal of solution

b) Setting up the problem inverted and multiplying by 1 000 000

c) Leaving out the specific gravity of the solution

Reference: Water Environment Federation (2008) *Operation of Municipal Wastewater Treatment Plants*, 6th ed.; Manual of Practice No. 11; Water Environment Federation: Alexandria, Virginia; pp 9-51–9-54.

8. **Answer: B**

Many of the common heavy metals form insoluble hydroxides at pH 11, so lime coagulation reduces these metal concentrations. Except for mercury, cadmium, and selenium, lime coagulation removes 90% or more of most heavy metals.

Reference: Water Environment Federation (2008) *Operation of Municipal Wastewater Treatment Plants*, 6th ed.; Manual of Practice No. 11; Water Environment Federation: Alexandria, Virginia; p 24-27.

9. **Answer: B**

Reference: Water Environment Federation (2008) *Operation of Municipal Wastewater Treatment Plants*, 6th ed.; Manual of Practice No. 11; Water Environment Federation: Alexandria, Virginia; p 32-42.

10. **Answer: C**

$$\text{Volatile solids reduction} = \frac{0.78 - 0.64}{0.78 - (0.78 \times 0.64)} \times 100 = 49.9 \text{ or } 50\%$$

Reference: Water Environment Federation (2008) *Operation of Municipal Wastewater Treatment Plants*, 6th ed.; Manual of Practice No. 11; Water Environment Federation: Alexandria, Virginia; p 30-5.

11. **Answer: A**

There are two groups of microorganisms in anaerobic digestion: the acid formers, or saprophytic bacteria, and the methane formers. The acid formers break down complex organics into acetic acid, propionic acid, hydrogen, carbon dioxide, and other simple organic acids. In the second stage, methane-forming bacteria convert acetate to methane and carbon dioxide.

In some texts, the anaerobic digestion process is shown as three stages, with the first stage consisting of enzymes produced by bacteria working outside of the bacterial cells to break complex organic material from primary and secondary sludge into soluble organic fatty acids, alcohols, carbon dioxide, and ammonia.

Reference: Water Environment Federation (2008) *Operation of Municipal Wastewater Treatment Plants*, 6th ed.; Manual of Practice No. 11; Water Environment Federation: Alexandria, Virginia; p 30-3.

12. **Answer:** A

For nitrogen removal, the sequencing batch reactor cycle time must be balanced between air-on and air-off cycle times. To decrease the ammonia concentration, the air-on time must be increased to give the nitrifying bacteria adequate time to react. Denitrification is relatively rapid. Generally speaking, the air-on time should be approximately two-thirds of the total react cycle and air-off time should be approximately one-third of the total react cycle.

Reference: Water Environment Federation (2008) *Operation of Municipal Wastewater Treatment Plants*, 6th ed.; Manual of Practice No. 11; Water Environment Federation: Alexandria, Virginia, pp 22-53 and 22-63.

13. **Answer:** C

A membrane bioreactor will generally produce a higher quality effluent than a conventional activated sludge facility. Because of the capability to retain higher mixed liquor concentrations and the fact that there is no need to construct secondary clarifiers, the MBR will have a much smaller footprint than a conventional activated sludge facility.

Reference: Water Environment Federation (2008) *Operation of Municipal Wastewater Treatment Plants*, 6th ed.; Manual of Practice No. 11; Water Environment Federation: Alexandria, Virginia; p 20-48.

14. **Answer:** B

The presence of clear supernatant above poorly settling sludge is often due to the presence of filamentous microorganisms. Because the sludge is settling slowly, it is able to capture fine particles on its way down, leaving a very clear supernatant. An overgrowth of filamentous microorganisms can be caused by a low food-to-microorganism ratio, low dissolved oxygen conditions, a nutrient deficiency, or low pH. The presence of filamentous organisms can be confirmed by examining the sludge under a microscope. Once confirmed, a review of operating conditions can help identify the cause and suggest corrective actions.

Reference: Water Environment Federation (2008) *Operation of Municipal Wastewater Treatment Plants*, 6th ed.; Manual of Practice No. 11; Water Environment Federation: Alexandria, Virginia; p 20-123.

15. **Answer:** D

The most common cause of digester foaming is organic overload, which results in the production of more volatile fatty acids than can be converted to methane. The acid formers (which release carbon dioxide) work much more quickly than the methane-forming microorganisms. The resulting increase in carbon dioxide typically increases foam formation.

Reference: Water Environment Federation (2008) *Operation of Municipal Wastewater Treatment Plants,* 6th ed.; Manual of Practice No. 11; Water Environment Federation: Alexandria, Virginia; p 30-69.

16. **Answer:** C

Mass, lb/d = (Flow, mgd)(Concentration, mg/L)(8.34 lb/mil. gal)

Mass, lb/d = (12.5 mgd)(320 mg/L)(8.34 lb/mil. gal)

Mass, lb/d = 33 360

If the primary clarifier is removing 35% of the influent BOD, then 100% − 35% or 65% is going on to the secondary treatment process.

(33 360 lb/d)(0.65 = 21 684)

In this example, the primary clarifier was able to remove 35% of the influent BOD. Primary clarifiers separate the readily settleable and floatable solids from wastewater. The percent removal that may be achieved at a particular treatment facility will depend on influent wastewater characteristics.

Reference: Formula sheet and Water Environment Federation (2008) *Operation of Municipal Wastewater Treatment Plants,* 6th ed.; Manual of Practice No. 11; Water Environment Federation: Alexandria, Virginia; p 19-2.

17. **Answer:** A

Digestion nearly ceases at approximately 10 °C (50 °F).

Reference: Water Environment Federation (2008) *Operation of Municipal Wastewater Treatment Plants,* 6th ed.; Manual of Practice No. 11; Water Environment Federation: Alexandria, Virginia; p 30-63.

Class IV Job Tasks

- Add chemicals to disinfect and deodorize water and other liquids (e.g., ammonia, chlorine, lime);
- Analyze laboratory data to evaluate and adjust processes;
- Follow industry safety rules and guidelines applicable to treatment processes;
- Implement changes as indicated by laboratory results;
- Operate chemical feed systems (e.g., polymer, ferric);
- Operate odor control systems (e.g., biofilters, scrubbers);

- Operate SCADA systems;
- Operate the preliminary treatment processes (e.g., screening, grit, flow equalization);
- Operate the primary clarification/sedimentation processes;
- Operate the following secondary treatment processes:
  - Secondary clarification/sedimentation processes and
  - Conventional activated sludge processes (e.g., step feed, plug flow, complete mix, MBR);
- Operate the following tertiary treatment processes:
  - Nutrient removal systems,
  - Filtration/ion exchange systems (e.g., sand, membranes), and
  - Filtration systems (e.g., solids, liquid); and
- Operate the disinfection processes (e.g., UV, ozone);
- Operate the following solids treatment processes:
  - Anaerobic digestion process,
  - Mechanical dewatering processes (e.g., presses, centrifuges), and
  - Solids thickening processes (e.g., DAF, belt, rotary drum).

Class IV Types of Knowledge Required to Perform Job Tasks

Types of Knowledge	Level of Knowledge
Aeration principles (e.g., mixing, mechanical, diffusers)	Advanced
Bacteriological laboratory testing (e.g., coliform, fecal, *E coli*)	Advanced
Biological laboratory testing (e.g., BOD, SOUR, CBOD)	Advanced
Biosolids disposal and monitoring requirements	Advanced
Chemical laboratory testing (e.g., ammonia, phosphorus, alkalinity)	Advanced
Chlorinators (e.g., gas, liquid)	Advanced
Clarifiers	Advanced
Comminuters	Advanced
Conveyors	Advanced
Dewatering equipment (e.g., centrifuges, presses, drying beds)	Advanced
Documentation and recordkeeping	Advanced
Effluent disposal and monitoring requirements	Advanced
Flow measuring devices (e.g., Parshall flumes, mag meter, venturis)	Advanced
Grit removal processes (e.g., grit chamber, aeration, cyclone)	Advanced
Hydraulic principles (e.g., mass flow balance, detention time, loading, velocity, HRT)	Advanced
Influent monitoring and waste characteristics	Advanced

Types of Knowledge	Level of Knowledge
Ozone generation equipment	Advanced
Physical laboratory testing (e.g., temperature, solids, dissolved oxygen)	Advanced
Pneumatic principles (e.g., troubleshooting actuators, compressors, sprayers)	Advanced
Primary treatment processes (e.g., ponds, sedimentation basins)	Advanced
Principles of asset management (e.g., preventive, reactive, predictive maintenance)	Advanced
Process control instrumentation (e.g., PLCs, SCADA, continuous online monitoring)	Advanced
Quality control/quality assurance practices	Advanced
Screening technology (e.g., bar screens, microscreens)	Advanced
Secondary treatment processes (e.g., activated sludge, MBR, SBR)	Advanced
Solids treatment concepts (e.g., dewatering, digestion, thickening)	Advanced
Tertiary treatment processes (e.g., media filtration, disinfection, post-aeration, reclaimed recharge)	Advanced
Treatment equipment (e.g., pumps, motors, generators)	Advanced
Wastewater treatment practices (e.g., sludge age, SRT, MCRT, F:M)	Advanced

Class IV Sample Questions

1. Activated sludge flows to secondary clarifiers. The clarifier blanket has increased over the last week from 1 to 4 ft. The most recent sludge volume index (SVI) is 175 mL/g. The most likely problem is

    a) overgrowth of filamentous organisms.
    b) too many clarifiers are online; take one out of service.
    c) short circuiting is causing upwelling at effluent weir.
    d) toxic substance has inhibited floc formers.

2. A considerable problem with complete-mix reactors is _____. One way to combat this problem is _____.

    a) Filamentous bulking / use a selector basin or zone in front of the process
    b) Hard to keep dissolved oxygen at 2 mg/L / add more aerators
    c) Return activated sludge (RAS) needs to be aerated / slow RAS rate
    d) High nitrates at the end of process / selector in front of process

3. You are the manager of an activated sludge facility with a low effluent nitrogen limit. In the last few days, you have slowly been increasing the chlorine dosage to compensate for increasing demand. The facility is currently in violation of the ammonia–nitrogen

limit of 10 mg/L as nitrogen and you are concerned that the chlorine residual is decreasing. The following has been documented in the facility logbook:

- The chlorine contact tanks were cleaned last week,
- Influent flow and loading have been normal,
- Temperatures have been steadily decreasing,
- Secondary effluent suspended solids loading is stable, and
- One of the aeration blowers failed in the last 3 weeks and it has been sent to the manufacturer.
- Dissolved oxygen concentrations cannot be maintained higher than 1 mg/L in the aeration basins.

The reduction in chlorine residual is most likely caused by _____ and could most economically be improved by _____.

a) Low effluent temperatures/applying heat to the effluent flow
b) Ammonia bleed-through and disinfection near the breakpoint/reducing the dosage in small increments
c) The formation of disinfection byproducts/adding ammonia
d) Low dissolved oxygen in the effluent/purchasing a surface aerator for the chlorine contact tank

4. Determine the number of chlorine containers that must be ordered for a 30-day period to maintain a sufficient supply of chlorine, given the following information:

- Average flow = 3.3 mgd (12 ML/d),
- Desired chlorine residual = 4 mg/L,
- Average chlorine demand = 6 mg/L, and
- Pounds of chlorine per ton container = 2000 lb (909.1 kg).

a) **One** 1-ton container
b) **Three** 1-ton containers
c) **Four** 1-ton containers
d) **Five** 1-ton containers

5. Calculate the annual power cost to operate the following UV disinfection system (below); assume all lamps are used 365 days per year.

- Power cost = $0.10/kWh
- Lamp power used = 700 W, and
- Number of lamps = 50

a) $1278 per year
b) $3500 per year

c) $30,660 per year

d) $306,600 per year

6. In the event of an anaerobic digester upset, which of the following is included in an effective troubleshooting strategy?

   a) Increase solids feed to the digester and monitor both alkalinity and volatile acids

   b) Monitor dissolved oxygen and burn off all excessive gas created

   c) Decrease mixing and monitor dissolved oxygen

   d) Decrease organic loading rate and monitor both alkalinity and volatile acids

7. The anaerobic digester supernatant is high in solids. Which of the following is the most probable cause?

   a) Excessive mixing and not enough settling time

   b) Supernatant draw-off point is at the same level as the supernatant layer

   c) Sludge feed point is located too far away from the supernatant draw-off line

   d) Withdrawing too much digested sludge

8. Moisture affects the rate of biological activity in the composting process. Which of the following is closest to the optimum moisture range for a composting process?

   a) 10–20% moisture

   b) 21–40% moisture

   c) 41–60% moisture

   d) 61–90% moisture

9. A tertiary filtration system is in danger of becoming hydraulically overloaded because of high levels of inflow/infiltration. Blending of filtered and nonfiltered secondary effluent is proposed. The filter capacity is 15 mgd (57 ML/d) and produces effluent with a suspended solids concentration of 4 mg/L. The current plant influent flow is 45 mgd (170 ML/d), and the secondary effluent was tested at 27 mg/L suspended solids. If the flows were blended, what would be the expected suspended solids concentration?

   a) 15.5 mg/L

   b) 19.3 mg/L

   c) 28.3 mg/L

   d) 30.0 mg/L

10. The dewatering centrifuge solids capture has been decreasing over the past week. An inspection of the unit shows that the pond depth is too low. What course of action should be taken?

   a) Conduct jar tests with diluted polymer, try adding post-dilution water to polymer, or thinning out the feed solids

   b) Increase the bowl speed by 15%

    c)   Check weir setting; raise pond level by 6 mm

    d)   Reduce torque setting to obtain dryer, more viscous cake

11. An activated sludge process has been operated as a high-rate activated sludge process. Which parameter needs to be adjusted to operate the process as a conventional activated sludge system?

    a)   Increase wasting to decrease mixed liquor suspended solids (MLSS)

    b)   Increase mean cell residence time (MCRT)

    c)   Increase the food-to-microorganism ratio

    d)   Remove an aeration basin from service

12. Denitrification under controlled conditions is used to remove nitrogen as nitrogen gas from wastewaters in anoxic reactors. However, when denitrification occurs in the sludge blanket of a secondary clarifier, it can lead to floating sludge that can increase the secondary effluent total suspended solids concentration. Which of the following statements correctly states an action an operator might take to minimize denitrification occurring in a secondary clarifier?

    a)   Decrease wasting to increase the solids retention time (SRT)

    b)   Put an additional secondary clarifier online to increase secondary clarifier detention time

    c)   Increase the return activated sludge (RAS) pumping rate to pull down the secondary clarifier sludge blanket and decrease the SRT if nitrification is not necessary

    d)   Decrease the food-to-microorganism ratio, regardless if nitrification is necessary

13. The mean cell residence time (MCRT) and the F:M are inversely related to each other. Which of the following statements about this relationship is true?

    a)   A low F:M filamentous bacteria could proliferate if the MCRT is too long.

    b)   Because a short MCRT is needed to nitrify, the F:M is high when a facility is nitrifying.

    c)   If the operator decreases the MCRT, the F:M would decrease.

    d)   Because it takes a minimum of 5 days to get the data needed to calculate the F:M, it is preferred over the MCRT for process control.

14. Many pond systems are equipped with an adjustable effluent weir that can be raised and lowered. How long would it take a pond to fill, ignoring precipitation, percolation and evaporation, if the operator raises the level in a 44-ac pond by 14 in.? During this period, there is no discharge from the pond and the influent flow averages 1.65 mgd.

    a)   10.1 days

    b)   31.1 days

    c)   95.5 days

    d)   122 days

15. What is the organic loading rate to a 550 000-gal anaerobic digester if the mixed primary and waste activated sludge flow averages 41.5 gpm and has a total solids content of 4.2%, 74% of which is volatile?

    a)   0.003 lb volatile solids (VS)/d·cu ft

    b)   0.028 lb VS/d·cu ft

    c)   0.21 lb VS/d·cu ft

    d)   21 lb VS/d·cu ft

Answers

1.   **Answer:** A

If an activated sludge is not settling or compacting well, and the sludge volume index is greater than 150 mL/g, chances are the problem is because of an overgrowth of filamentous organisms.

Reference: Water Environment Federation (2008) *Operation of Municipal Wastewater Treatment Plants*, 6th ed.; Manual of Practice No. 11; Water Environment Federation: Alexandria, Virginia; p 44-46.

2.   **Answer:** A

A selector that is either anaerobic or anoxic can be set up in front of the complete-mix reactor. The advantage to this is filamentous organisms need oxygen to multiply and non-filamentous organisms do not.

Reference: Water Environment Federation (2008) *Operation of Municipal Wastewater Treatment Plants*, 6th ed.; Manual of Practice No. 11; Water Environment Federation: Alexandria, Virginia; p 20-51.

3.   **Answer:** B

The facility is not nitrifying completely because of the loss of an aeration blower and the nitrogen limit is being exceeded. It would appear that the chlorination system is experiencing operation on the chloramine destruction phase of the breakpoint curve. This is evident because of the residual reduction with an increase in dosage. Small incremental decreasing of chlorine would theoretically move the process to the chloramine formation portion of the curve and increase the residual in the most economical way. Previously, because ammonia concentrations were low, the effluent contained free chlorine residual. As effluent ammonia concentrations increased, it shifted the chlorine-to-ammonia ratio below that required to achieve breakpoint chlorination. Further reducing the effluent chlorine dose will likely increase the combined chlorine residual, while simultaneously reducing operating costs.

Reference: Water Environment Federation (2008) *Operation of Municipal Wastewater Treatment Plants*, 6th ed.; Manual of Practice No. 11; Water Environment Federation: Alexandria, Virginia; p 26-29.

4.   **Answer: D**

Solution:

When chlorine is added to water, some of the chlorine is consumed and some will combine with organic compounds and ammonia in the water. The amount of chlorine added (dose) will always be higher than the amount of chlorine that can be measured (residual) because of the demand. This is shown mathematically as

Dose = Demand + Residual
Dose = 6 mg/L + 4 mg/L
Dose = 10 mg/L

Use the feed rate equation. Because this is pure chlorine, the purity is 100% or just 1.

$$\text{Feed rate, lb/d} = \frac{(\text{Dosage, mg/L})(\text{Capacity, mgd})(8.34 \text{ lb/mil. gal})}{\text{Purity, \% expressed as a decimal}}$$

Feed rate, lb/d = (10 mg/L)(3.3 mgd)(8.34 lb/mil. gal)
Feed rate, lb/d = 275.22

Now find the amount needed for 30 days:

$$\frac{30 \text{ days}}{} \left[\frac{275.22 \text{ lb}}{1 \text{ day}} \right] \left[\frac{1 \text{ container}}{2000 \text{ lb}} \right] = 4.12 \text{ containers}$$

Because you need more than 4 containers, order 5.

Chlorine is commonly shipped in 150-lb cylinders, 1-ton containers, and rail cars.

Reference: Water Environment Federation (2008) *Operation of Municipal Wastewater Treatment Plants*, 6th ed.; Manual of Practice No. 11; Water Environment Federation: Alexandria, Virginia; pp 9-13, 9-51, and 26-30.

5.   **Answer: B**

700 W × 0.001 kW/W × 50 lamps × 24 hr/d × $0.1/kWh × 365 d/yr = $30,660/yr

Reference: Water Environment Federation (2008) *Operation of Municipal Wastewater Treatment Plants*, 6th ed.; Manual of Practice No. 11; Water Environment Federation: Alexandria, Virginia; p 22-53.

Reference: California State University, Sacramento (2007) *Operation of Wastewater Treatment Plants*, 7th ed.; California State University: Sacramento, California; Volume II, p 376.

6.   **Answer: D**

In a properly operating anaerobic digester, the bicarbonate alkalinity should be maintained at a level no lower than 1000 mg/L as calcium carbonate to ensure adequate pH control. In an upset condition, net buffer [alkalinity] consumption takes place and the process is in danger of pH failure. The volatile acids-to-alkalinity ratio should be maintained below 0.2 to prevent souring of the digester.

Reference: Water Environment Federation (2008) *Operation of Municipal Wastewater Treatment Plants*, 6th ed.; Manual of Practice No. 11; Water Environment Federation: Alexandria, Virginia; pp 25-44, 30-64, and Table 30.7 on p 30-54.

7. **Answer:** A

Allow longer settling periods before withdrawing supernatant. Fill a 10- to 20-L glass carboy and observe the separation pattern.

Reference: Water Environment Federation (2008) *Operation of Municipal Wastewater Treatment Plants*, 6th ed.; Manual of Practice No. 11; Water Environment Federation: Alexandria, Virginia; p 30-77.

8. **Answer:** C

Moisture affects the rate of biological activity. At less than approximately 40% moisture, activity begins to decrease. At approximately 60% moisture, the air pore space is blocked. This affects the aeration efficiency of the system and results in anaerobic zones in the compost bed.

Reference: Water Environment Federation (2008) *Operation of Municipal Wastewater Treatment Plants*, 6th ed.; Manual of Practice No. 11; Water Environment Federation: Alexandria, Virginia; p 32-11.

9. **Answer:** B

The mixing equation is as follows:

$$c_3 = \frac{Q_1 c_1 + Q_2 c_2}{Q_1 + q_2}$$

c_3 = mixed concentration,
c_1 = concentration at point 1 = 4 mg/L,
c_2 = concentration at point 2 = 27 mg/L
Q_1 = flow at point 1 = 15 mgd, Q_2 = flow at point 2 = 45 mgd – 15 mgd = 30 mgd

$$c_3 = \frac{15 \text{ mgd} \times 4 \text{ mg/L} + 30 \text{ mgd} \times 27 \text{ mg/L}}{15 \text{ mgd} + 30 \text{ mgd}} = 19.3 \text{ mg/L}$$

10. **Answer:** C

Thickening centrifuges react to very small changes in pond height. A fair-sized change in a pond is 1.5 mm (1/16 in.). The goal is to set the pond so that good operation occurs with the differential revolutions per minute somewhere in its mid-range. Changing the feed rate substantially will require a change in pond setting.

Considering the pond surface (the liquid–air interface within the centrifuge), raising the pond surface reduces pond radius and thus reduces the surface area of the pond where separation takes place. This should give better centrate quality.

Reference: Water Environment Federation (2008) *Operation of Municipal Wastewater Treatment Plants*, 6th ed.; Manual of Practice No. 11; Water Environment Federation: Alexandria, Virginia; pp 29-65 and 33-82.

11. Answer: B

High-rate activated sludge systems operate at shorter MCRTs and higher food-to-microorganism ratios (F:Ms). An activated sludge process that is being operated under conventional loading rates will have an MCRT between 5 and 15 days and an (F:M) between 0.2 and 0.5 kg/kg·d (lb/lb·d).

The process control variables of MCRT, MLSS concentration, F:M, and wasting rate are all related to one another. It is not possible to change one of these variables without causing the others to also change. When MCRT is decreased, this results in the need for a higher wasting rate, which will decrease the MLSS concentration and increase F:M.

Table 20.1   Typical process loading ranges for the activated sludge process.

Loading range	MCRT, d	Volumetric loading, (lb BOD/1000 cu ft) kg BOD/m³	F:M, kg/kg·d/ (lb/lb·d)
High rate	1–3	1.60–16.0 (0.5–1.5)	0.5–1.5 (100–1000)
Conventional	5–15	0.32–0.64 (0.2–0.5)	0.2–0.5 (20–40)
Low rate	20–30	0.16–0.40 (0.05–01.5)	0.05–0.15 (10–25)

Reference: Water Environment Federation (2008) *Operation of Municipal Wastewater Treatment Plants*, 6th ed.; Manual of Practice No. 11; Water Environment Federation: Alexandria, Virginia; pp 20-12, 20-41.

12. Answer: C

For denitrification to occur in the secondary clarifier, nitrate must be present. Facultative bacteria convert nitrate to nitrogen gas in the absence of dissolved oxygen. Preventing denitrification requires that either the formation of nitrate be prevented in the first place by preventing nitrification (the conversion of ammonia to nitrate) in the secondary treatment process or by removing the settled mixed liquor suspended solids from the clarifier before it has an opportunity to complete the conversion of nitrate to nitrogen gas. To prevent nitrate formation, the operator may decrease sludge age by increasing the waste activated sludge rate by not more than 10% per day to reduce or eliminate nitrification. To prevent nitrate usage in the clarifier blanket, adjust RAS rates to maintain sludge blanket depths below 0.3 to 0.9 m (1 to 3 ft).

Reference: Water Environment Federation (2008) *Operation of Municipal Wastewater Treatment Plants,* 6th ed.; Manual of Practice No. 11; Water Environment Federation: Alexandria, Virginia; pp 20-55–20-58 and Table 20.16, p 20-165.

13. Answer: A

A longer MCRT leads to more solids being kept in the system and an increase in mixed liquor suspended solids concentration. If the same amount of food is entering the facility on average, then the F:M will necessarily decrease. Filamentous bacteria that proliferate under low F:M conditions may also proliferate at a longer MCRT.

Reference: Water Environment Federation (2008) *Operation of Municipal Wastewater Treatment Plants,* 6th ed.; Manual of Practice No. 11; Water Environment Federation: Alexandria, Virginia; pp 20-39 and 20-40.

14. Answer: A

Use the detention time formula:

$$\text{Detention Time} = \frac{\text{Volume}}{\text{Flow}}$$

First, find the volume available. Because the influent flow is in millions of gallons, the volume must be in million gallons. Start by finding dimensions in feet:

$$44 \text{ ac} \left[\frac{43\,560 \text{ sq ft}}{1 \text{ ac}} \right] = 1\,916\,640 \text{ sq ft}$$

$$14 \text{ in.} \left[\frac{1 \text{ ft}}{12 \text{ in.}} \right] = 1.167 \text{ ft}$$

Volume = (Length)(Width)(Height)
Volume = (Area)(Height)
Volume = (1 916 640 sq ft)(1.167 ft)
Volume = 2 236 719 cu ft

$$2\,236\,7169 \text{ cu ft} \left[\frac{7.48 \text{ gal}}{1 \text{ cu ft}} \right] \left[\frac{1 \text{ mil. gal}}{1\,000\,000 \text{ gal}} \right] = 16.7 \text{ mil. gal}$$

Now that the units go together, use the detention time formula:

$$\text{Detention Time} = \frac{\text{Volume}}{\text{Flow}}$$

$$\text{Detention Time} = \frac{16.7 \text{ mil. gal}}{1.65 \text{ mgd}}$$

$$\text{Detention Time} = 10.1 \text{ days}$$

15. **Answer: C**

First, find the total mass of solids entering the digester.

Convert influent percent solids to mg/L:

$$4.2\% \left[\frac{10\ 000 \text{ mg/L}}{1\%} \right] = 42\ 000 \text{ mg/L}$$

$$\frac{41.5 \text{ gal}}{\text{min}} \left[\frac{1 \text{ mil. gal}}{1\ 000\ 000 \text{ gal}} \right] \left[\frac{60 \text{ min}}{1 \text{ hour}} \right] \left[\frac{24 \text{ hours}}{1 \text{ day}} \right] = 0.059\ 76 \text{ mgd}$$

Mass, lb/d = (Volume, mil. gal)(Concentration, mg/L)(8.34 lb/mil. gal)

Mass, lb/d = (0.059 76 mil. gal)(42 000 mg/L)(8.34 lb/mil. gal)

Mass, lb/d = 20 932.7

Only a portion of the total solids are volatile. Find the volatile solids:

$$20\ 932.7 \text{ lb total solids} \left[\frac{74 \text{ lb volatile solids}}{100 \text{ lb total solids}} \right] = 15\ 490 \text{ lb}$$

Calculate the solids loading rate:

$$550\ 000 \text{ gal} \left[\frac{1 \text{ cu ft}}{7.48 \text{ gal}} \right] = 73\ 529.4 \text{ cu ft}$$

$$\text{Solids Loading Rate, lb/d/cu ft} = \frac{\text{Mass of Volatile Solids, lb}}{\text{Digester Volume, cu ft}}$$

$$\text{Solids Loading Rate, lb/d/cu ft} = \frac{15\ 490 \text{ lb volatile solids}}{73\ 529.4 \text{ cu ft}}$$

$$\text{Solids Loading Rate, lb/d/cu ft} = 0.21$$

Laboratory Analysis

Class I Job Tasks

- Follow laboratory standard operating procedures (SOPs);
- Collect samples for the following:
  - Bacteriological analyses,
  - Biological analyses (e.g., BOD, CBOD),
  - Chemical analyses (e.g., COD, nutrients, metals), and
  - Physical analyses (e.g., pH, temperature, dissolved oxygen, settleable solids);
- Conduct the following:
  - Physical analyses and
  - Process control laboratory testing; and
- Interpret data from the following:
  - Bacteriological analyses,
  - Biological analyses (e.g., BOD, CBOD),
  - Chemical analyses (e.g., COD, nutrients, metals), and
  - Physical analyses (e.g., pH, temperature, dissolved oxygen, settleable solids).

Class I Types of Knowledge Required to Perform Job Tasks

Types of Knowledge	Level of Knowledge
Bacteriological laboratory testing (e.g., coliform, fecal, *E coli*)	Basic
Biological laboratory testing (e.g., BOD, SOUR, CBOD)	Basic
Biosolids disposal and monitoring requirements	Basic
Chemical laboratory testing (e.g., ammonia, phosphorus, alkalinity)	Basic
Documentation and recordkeeping	Basic
Effluent disposal and monitoring requirements	Basic
Industry safety practices (e.g., PPE, confined spaces, fall arrest, lockout/tagout)	Basic
Influent monitoring and waste characteristics	Basic
Physical laboratory testing (e.g., temperature, solids, dissolved oxygen)	Basic
Process control instrumentation (e.g., PLCs, SCADA, continuous online monitoring)	Basic
Quality control/quality assurance practices	Basic
Wastewater treatment practices (e.g., sludge age, SRT, MCRT, F:M)	Basic

Class I Sample Questions

1.   Which is the standardized condition for biochemical oxygen demand (BOD) analysis?

   a)   20 °C for 3 days
   b)   20 °C for 5 days
   c)   25 °C for 5 days
   d)   25 °C for 3 days

2.   Which statement is true regarding suspended solids in activated sludge?

   a)   Volatile suspended solids represent the inorganic fraction in mixed liquor.
   b)   Total suspended solids analysis requires filtering, while total volatile suspended solids analysis does not.
   c)   There are no inert materials in mixed liquor volatile suspended solids (MLVSS).
   d)   Particulate matter in aeration tanks is called "mixed liquor suspended solids" (MLSS).

Answers

1.   **Answer: B**

   Because the rate of biological activity depends on temperature and complete stabilization as long as 20 days may be required, the 5-day BOD test has been standardized to conditions of 20 °C for 5 days to ensure consistent results between all laboratories. These conditions will typically provide a value equivalent to the carbonaceous oxygen demand. However, samples that include nitrifiers may need to be treated with an inhibitor to avoid including nitrogenous oxygen demand. Because the rate of biological activity depends on temperature and complete stabilization, as long as 20 days may be required.

   Reference: Water Environment Federation (2008) *Operation of Municipal Wastewater Treatment Plants*, 6th ed.; Manual of Practice No. 11; Water Environment Federation: Alexandria, Virginia; p 17-12.

2.   **Answer: D**

   Particulate matter is referred to as "mixed liquor suspended solids" (MLSS) and the organic fraction is called "mixed liquor volatile suspended solids" (MLVSS).

   Reference: Water Environment Federation (2008) *Operation of Municipal Wastewater Treatment Plants*, 6th ed.; Manual of Practice No. 11; Water Environment Federation, Alexandria, Virginia; p 20-3.

Class II Job Tasks

- Follow laboratory standard operating procedures (SOPs);
- Collect samples for the following:
  - Bacteriological analyses,
  - Biological analyses (e.g., BOD, CBOD),
  - Chemical analyses (e.g., COD, nutrients, metals), and
  - Physical analyses (e.g., pH, temperature, dissolved oxygen, settleable solids);
- Conduct the following:
  - Bacteriological analyses,
  - Biological analyses (e.g., BOD, CBOD),
  - Chemical analyses (e.g., COD, nutrients, metals),
  - Physical analyses,
  - Process control laboratory testing, and
  - Required regulatory laboratory testing; and
- Interpret data from the following:
  - Bacteriological analyses,
  - Biological analyses (e.g., BOD, CBOD),
  - Chemical analyses (e.g., COD, nutrients, metals), and
  - Physical analyses (e.g., pH, temperature, dissolved oxygen, settleable solids).

Class II Types of Knowledge Required to Perform Job Tasks

Types of Knowledge	Level of Knowledge
Bacteriological laboratory testing (e.g., coliform, fecal, *E coli*)	Intermediate
Biological laboratory testing (e.g., BOD, SOUR, CBOD)	Intermediate
Biosolids disposal and monitoring requirements	Basic
Chemical laboratory testing (e.g., ammonia, phosphorus, alkalinity)	Intermediate
Documentation and recordkeeping	Basic
Effluent disposal and monitoring requirements	Intermediate
Industry safety practices (e.g., PPE, confined spaces, fall arrest, lockout/tagout)	Basic
Influent monitoring and waste characteristics	Basic
Physical laboratory testing (e.g., temperature, solids, dissolved oxygen)	Intermediate
Process control instrumentation (e.g., PLCs, SCADA, continuous online monitoring)	Basic
Quality control/quality assurance practices	Basic
Wastewater treatment practices (e.g., sludge age, SRT, MCRT, F:M)	Intermediate

Class II Sample Questions

1. Which of the following is used to measure settleable solids in primary clarifier influents and effluents?

   a) Imhoff cone
   b) Gooch crucible
   c) Buchner funnel
   d) Reflux column

2. Which statement is true about pH?

   a) The pH scale ranges from 0 to 14.
   b) Neutral pH represents a reading of 7.
   c) pH above 7 indicates acidic condition.
   d) pH below 7 indicates a basic condition.

3. A 50-mL sample has been filtered through a prepared Gooch crucible. The crucible and filtered solids have been dried at 103 °C for 1 hour. The dried weight is 12.3940 g and the tare weight was 12.3795 g. What is the total suspended solids concentration in the sample?

   a) 55 mg/L
   b) 145 mg/L
   c) 290 mg/L
   d) 550 mg/L

4. When measuring fecal coliform bacteria using the membrane filtration method, the sample volume should be adjusted as needed so at least one plate will have what range of colonies?

   a) 1–40 with a maximum of 100
   b) 20–60 with a maximum of 200
   c) 40–80 with a maximum of 300
   d) 60–100 with a maximum of 400

5. The correlation between carbonaceous biochemical oxygen demand (cBOD) and chemical oxygen demand (COD) varies from facility to facility; however, for facilities receiving primarily domestic wastewater, the COD-to-BOD ratio for raw, unsettled water is typically

   a) 0.5 to 1
   b) 1.9 to 2.2
   c) 5.2 to 1
   d) 10 to 1

Answers

1. **Answer:** A

   The settleable solids test can be performed to show quickly and qualitatively if the primary and secondary settling processes are functioning properly. The Imhoff cone represents ideal settling conditions. If the Imhoff cone removes more material than the primary or secondary treatment process being evaluated, then treatment process efficiency could be improved. An Imhoff cone looks like a large centrifuge tube that holds 1 L and has graduations on the tip of the cone and are read as milliliters of settled solids per liter of sample.

   Reference: Water Environment Federation (2012) *Basic Laboratory Procedures for the Operator–Analyst*, 5th ed.; Water Environment Federation: Alexandria, Virginia; pp 109–111.

2. **Answer:** B

   The pH scale ranges from 0 to 14, with a neutral reading of 7. Readings below 7 indicate an acidic condition and those above 7 indicate a basic condition.

   Reference: Water Environment Federation (2008) *Operation of Municipal Wastewater Treatment Plants*, 6th ed.; Manual of Practice No. 11; Water Environment Federation: Alexandria, Virginia; p 17-11.

3. **Answer:** C

   $$\text{Solids, mg/L} = \frac{(\text{Dry Solids, g})(1000 \text{ mg/g})(1000 \text{ mL/L})}{\text{Sample Volume, mL}}$$

   $$\text{Solids, mg/L} = \frac{(12.3940 \text{ g} - 12.3795 \text{ g})(1000 \text{ mg/g})(1000 \text{ mL/L})}{50 \text{ mL}}$$

   $$\text{Solids, mg/L} = \frac{(0.0145 \text{ g})(1000 \text{ mg/g})(1000 \text{ mL/L})}{50 \text{ mL}}$$

   $$\text{Solids, mg/L} = 290$$

   Reference: Water Environment Federation (2012) *Basic Laboratory Procedures for the Operator–Analyst*, 5th ed.; Water Environment Federation: Alexandria, Virginia; pp 126–139.

4. **Answer:** B

   The desired colony count to be used in calculating the number of colony forming units per 100 mL is 20 to 60 colonies per plate, with no more than 200 total colonies on the plate. Sample volumes should be adjusted to achieve this range when possible.

   Reference: Water Environment Federation (2012) *Basic Laboratory Procedures for the Operator–Analyst*, 5th ed.; Water Environment Federation: Alexandria, Virginia; pp 307–308.

5.   **Answer:** B

Chemical oxygen demand (COD) and biochemical oxygen demand (BOD) are related to one another. The COD test measures all of the substances in wastewater that may consume oxygen under the right conditions. This includes organic compounds that can be eaten by microorganisms in the treatment process; organic compounds that cannot be eaten by microorganisms during the 5-day BOD test; and inorganic compounds such as nitrite, ferrous iron, sulfide, manganous manganese, and other compounds. The BOD test only measures those organic substances that can be consumed by microorganisms within 5 days. Because of this, COD will always be equal to or larger than BOD for a particular wastewater sample. For domestic influent wastewater, the COD concentration for a raw, unsettled sample will be between 1.9 and 2.2 times the cBOD concentration.

Reference: Water Environment Federation (2008) *Operation of Municipal Wastewater Treatment Plants,* 6th ed.; Manual of Practice No. 11; Water Environment Federation: Alexandria, Virginia; p 17-19.

Class III Job Tasks

- Follow laboratory standard operating procedures (SOPs);
- Collect samples for the following:
    - Bacteriological analyses,
    - Biological analyses (e.g., BOD, CBOD),
    - Chemical analyses (e.g., COD, nutrients, metals), and
    - Physical analyses (e.g., pH, temperature, dissolved oxygen, settleable solids);
- Conduct the following:
    - Bacteriological analyses,
    - Biological analyses (e.g., BOD, CBOD),
    - Chemical analyses (e.g., COD, nutrients, metals),
    - Physical analyses,
    - Process control laboratory testing, and
    - Required regulatory laboratory testing; and
- Interpret data from the following:
    - Bacteriological analyses,
    - Biological analyses (e.g., BOD, CBOD),
    - Chemical analyses (e.g., COD, nutrients, metals), and
    - Physical analyses (e.g., pH, temperature, dissolved oxygen, settleable solids).

Class III Types of Knowledge Required to Perform Job Tasks

Types of Knowledge	Level of Knowledge
Bacteriological laboratory testing (e.g., coliform, fecal, *E coli*)	Intermediate
Biological laboratory testing (e.g., BOD, SOUR, CBOD)	Intermediate
Biosolids disposal and monitoring requirements	Intermediate
Chemical laboratory testing (e.g., ammonia, phosphorus, alkalinity)	Intermediate
Documentation and recordkeeping	Intermediate
Effluent disposal and monitoring requirements	Intermediate
Industry safety practices (e.g., PPE, confined space, fall arrest, lockout/tagout)	Intermediate
Influent monitoring and waste characteristics	Intermediate
Physical laboratory testing (e.g., temperature, solids, dissolved oxygen)	Advanced
Process control instrumentation (e.g., PLCs, SCADA, continuous online monitoring)	Intermediate
Quality control/quality assurance practices	Intermediate
Wastewater treatment practices (e.g., sludge age, SRT, MCRT, F:M)	Advanced

Class III Sample Questions

1. Which statement is correct regarding total and fecal coliform tests?

   a) Total and fecal coliform bacteria are pathogens.
   b) Both total and fecal coliform bacteria are reported as colonies per 10 mL.
   c) Total and fecal coliform bacteria analysis must be processed immediately.
   d) Total and fecal coliform bacteria are resistant to disinfection.

2. A total alkalinity test is performed on a digested sludge sample. The sample volume is 50 mL. A total of 46 mL of the 0.1 N acid titrant is used to reach the endpoint. What is the total alkalinity of the sample in milligrams per liter as calcium carbonate ($CaCO_3$)?

   a) 460 mg/L
   b) 920 mg/L
   c) 2300 mg/L
   d) 4600 mg/L

3. Nitrogen is typically present in raw, domestic wastewater in all of these forms EXCEPT

   a) organic nitrogen.
   b) ammonia nitrogen.

c) total Kjeldahl nitrogen.

d) nitrite nitrogen.

Answers

1. **Answer:** D

    Total and fecal coliform bacteria are not pathogens themselves; instead, they are used as indicator organisms because they tend to resist the effects of disinfection better than most pathogens.

    Reference: Water Environment Federation (2008) *Operation of Municipal Wastewater Treatment Plants*, 6th ed.; Manual of Practice No. 11; Water Environment Federation: Alexandria, Virginia; p 17-13.

2. **Answer:** D

    $$\text{Alakalinity, mg/L as CaCO}_3 = \frac{(\text{Titrant Volume, mL})(\text{Acid Normality})(50\ 000)}{\text{Sample Volume, mL}}$$

    $$\text{Alakalinity, mg/L as CaCO}_3 = \frac{(46\ \text{mL})(0.1\ \text{N})(50\ 000)}{50\ \text{mL}}$$

    $$\text{Alakalinity, mg/L as CaCO}_3 = 4600$$

    Reference: Water Environment Federation (2012) *Basic Laboratory Procedures for the Operator–Analyst*, 5th ed.; Water Environment Federation: Alexandria, Virginia; p 99.

3. **Answer:** D

    Raw wastewater contains little or no nitrite and nitrate. Facultative bacteria in the collection system convert nitrite and nitrate to nitrogen gas when dissolved oxygen is not present.

    Reference: Water Environment Federation (2012) *Basic Laboratory Procedures for the Operator–Analyst*, 5th ed.; Water Environment Federation: Alexandria, Virginia; pp 9 and 10.

Class IV Job Tasks

- Follow laboratory standard operating procedures (SOPs);
- Collect samples for the following:
  - Bacteriological analyses,
  - Biological analyses (e.g., BOD, CBOD),
  - Chemical analyses (e.g., COD, nutrients, metals), and
  - Physical analyses (e.g., pH, temperature, dissolved oxygen, settleable solids);

- Conduct the following:
  - Bacteriological analyses,
  - Biological analyses (e.g., BOD, CBOD),
  - Chemical analyses (e.g., COD, nutrients, metals),
  - Physical analyses,
  - Process control laboratory testing, and
  - Required regulatory laboratory testing; and
- Interpret data from the following:
  - Bacteriological analyses,
  - Biological analyses (e.g., BOD, CBOD),
  - Chemical analyses (e.g., COD, nutrients, metals), and
  - Physical analyses (e.g., pH, temperature, dissolved oxygen, settleable solids).

Class IV Types of Knowledge Required to Perform Job Tasks

Types of Knowledge	Level of Knowledge
Bacteriological laboratory testing (e.g., coliform, fecal, *E coli*)	Advanced
Biological laboratory testing (e.g., BOD, SOUR, CBOD)	Advanced
Biosolids disposal and monitoring requirements	Advanced
Chemical laboratory testing (e.g., ammonia, phosphorus, alkalinity)	Advanced
Documentation and recordkeeping	Advanced
Effluent disposal and monitoring requirements	Advanced
Industry safety practices (e.g., PPE, confined spaces, fall arrest, lockout/tagout)	Advanced
Influent monitoring and waste characteristics	Advanced
Physical laboratory testing (e.g., temperature, solids, dissolved oxygen)	Advanced
Process control instrumentation (e.g., PLCs, SCADA, continuous online monitoring)	Advanced
Quality control/quality assurance practices	Advanced
Wastewater treatment practices (e.g., sludge age, SRT, MCRT, F:M)	Advanced

Class IV Sample Questions

1. What is a modified biochemical oxygen demand (BOD) test approved by the U.S. Environmental Protection Agency (U.S. EPA)?

   a) A BOD test that gives the result within 3 hours

   b) A BOD test that represents the combination of carbonaceous BOD (cBOD) and nitrogenous BOD (NBOD) values

     c)   A BOD test that adds a nitrification inhibitor for obtaining solely cBOD

     d)   A BOD test that is done by titration

2.   A 200-mL/L value of settled sludge volume (SSV) and 2000 mg/L of mixed liquor suspended solids (MLSS) concentrations of an aeration tank sample are obtained from laboratory experiments. What is the sludge volume index (SVI) of this sample?

     a)   10 mL/g

     b)   40 mL/g

     c)   50 mL/g

     d)   100 mL/g

3.   A sample of raw wastewater is analyzed for biochemical oxygen demand (BOD). A 5-mL sample is used in a standard 300 mL BOD bottle. No seed is added. The initial dissolved oxygen is measured at 7.5 mg/L. After a 5-day incubation period, the dissolved oxygen is 1.4 mg/L. What is the BOD of the sample?

     a)   36 mg/L

     b)   72 mg/L

     c)   366 mg/L

     d)   400 mg/L

Answers

1.   **Answer: C**

Because nitrogenous oxidation can sometimes result in substantially higher 5-day BOD (BOD_5) test results than those from oxidation of organic substances, U.S. EPA has approved the use of a modified BOD test in which a nitrification inhibitor is used to suppress the nitrification reaction. Use of the CBOD test for permit reporting must be approved by each state regulatory agency.

Reference: Water Environment Federation (2008) *Operation of Municipal Wastewater Treatment Plants*, 6th ed.; Manual of Practice No. 11; Water Environment Federation: Alexandria, Virginia; p 17-13.

2.   **Answer: D**

The sludge volume index is defined as follows:

$$\text{Sludge Volume Index, mL/g} = \frac{(\text{Settled Sludge Volume at 30 min})(1000 \text{ mg/g})}{\text{MLSS, mg/L}}$$

$$\text{Sludge Volume Index, mL/g} = \frac{(200 \text{ mL})(1000 \text{ mg/g})}{2000 \text{ mg/L}}$$

$$\text{Sludge Volume Index, mL/g} = 100$$

The amount of space in mL occupied by 1 g of solids after settling for 30 minutes.

The units are milliliters per gram. 2000 mg/L of mixed liquor suspended solids = 2000 mg/1000 mg/g = 2 g

Therefore 200 mL/2 g = 100 mL/g.

Reference: Water Environment Federation (2008) *Operation of Municipal Wastewater Treatment Plants*, 6th ed.; Manual of Practice No. 11; Water Environment Federation: Alexandria, Virginia; p 17-18.

3. **Answer: C**

$$\text{BOD, mg/L} = \frac{[(\text{Initial Dissolved Oxygen, mg/L}) - (\text{Final Dissolved Oxygen, mg/L})](300 \text{ mL})}{\text{Sample Volume, mL}}$$

$$\text{BOD, mg/L} = \frac{[(7.5 \text{ mg/L} - 1.4 \text{ mg/L})](300 \text{ mL})}{5 \text{ mL}}$$

$$\text{BOD, mg/L} = \frac{(6.1 \text{ mg/L})(300 \text{ mL})}{5 \text{ mL}}$$

$$\text{BOD, mg/L} = 366$$

Reference: Water Environment Federation (2012) *Basic Laboratory Procedures for the Operator–Analyst*, 5th ed.; Water Environment Federation: Alexandria, Virginia; p 180.

www.ingramcontent.com/pod-product-compliance
Lightning Source LLC
LaVergne TN
LVHW070229211224
799324LV00003BB/46